STEAM SHEDS
AND THEIR LOCOMOTIVES

C. J. Gammell

Publishing

Contents

Previous page:
BR Standard Class 2MT No 84005 (14E) awaits the right of way at Cardington on 14 December 1961. *J. Phillips*

Below:
Ivatt '2MT' No 46401 slumbers on Builth Wells shed, a sub shed of Shrewsbury (84G). *J. Phillips*

Right:
Preserved Stanier Class 5 No 45407 stands outside Carnforth shed (24L, later 10A) on 21 June 1975. *Author*

Front cover, top:
A view of Tebay shed (11D). *J. Phillips*

Front cover, bottom:
The nameplate of 'Jubilee' class 4-6-0 Leander is seen being cleaned at Shildon in August 1975. *Author*

Back cover, top:
'Princess Coronation' class 4-6-2 No 46246 City of Manchester seen in ex-works condition at Shrewsbury in 1960. *Author*

Back cover, bottom:
A close-up of the nameplate of 'Merchant Navy' class 4-6-2 No 35004 Cunard White Star. *J. Phillips*

Back cover inset:
King's Cross (34A) shed plate. *J. F. Aylard*

First published 1995

ISBN 978-0-7110-2395-6

© Chris Gammell 1995

Designed by Alan C. Butcher

Published by Ian Allan Publishing

an imprint of Ian Allan Ltd, Terminal House, Station Approach, Shepperton, Surrey TW17 8AS.

This edition produced by TAJ Books - 2007
Printed in China

Introduction

The great steam age was an awe-inspiring era in Britain during the 1950s and 1960s. Upon Nationalisation in 1948 British Railways could boast a collection of nearly 20,000 locomotives of every conceivable type, shape and colour. With over 400 classes there was plenty of variety. From 1951 onwards BR introduced Standard locomotives and these could be found all over the system. The locomotive policy changed rapidly during the late 1950s and a quick rundown of steam power was orchestrated so that by August 1968, 20 years after Nationalisation, the last steam train had run.

Looking back to those steamy years in Britain reminds us of that great heritage that passed so quickly. The main lines thrived with high-speed trains hauled by sleek Pacifics. In contrast, the humblest goods train could be seen pottering along hauled by something out of the 19th century. Many of the old company types survived into the BR era and pre-1923 engines were mixed in with the more modern classes. Engine sheds were not very healthy places and soot, grime and ashes mixed with coal smoke to form a cocktail of sulphurous smoke that was acrid to both the throat and nostrils. Repairs to engine sheds were a low priority for the prewar companies. When repair and maintenance was required, the fitters often found they had very dark places to work in. Shed staff often had to work out in the open in poor light and bad weather. Only the Midland, North Eastern and Great Western companies went in for enclosed roundhouses and even these trapped the release of smoke from stabled engines. Many engine sheds were bombed during wartime as they were easy targets from the air. BR was slow to repair the damage in the postwar years but some engine sheds were rebuilt during the late 1950s. Some steam sheds were reconstructed with the intention of conversion to the new form of traction. Some became diesel maintenance depots, a number of which are still in use.

The updating of BR heralded in the Modernisation Plan of 1955 took some years to undertake and did not start in earnest until the late 1950s. For the period under review, the Ian Allan Locomotives/Locoshed book for 1959 has been used as a reference to shed allocations. The Paul Bolger books published by Ian Allan have also been referred to and are a valuable source of reference. The year 1959 was the best year to pick as main line dieselisation had not got into full swing and the travelling public was still able to witness main line express steam on most lines. In this book each shed group has been shown, starting with the principal shed and then working down to the minor establishments in the group, starting with the London Midland Region. The LMS shedcode system is used, having been adopted on the whole of BR from 1950. The LMS inherited the system from the former Midland Railway. Shedcodes changed from 1950 onwards and with the rundown of steam districts and divisions the Regions changed their boundaries particularly on the LMR and NER where regional boundaries were changed on more than one occasion. Sample allocations for different sheds have been shown, mostly for 1959 but some 1950 allocations have been included.

In this book the date of closure of a certain depot shown is the date that steam ceased to be used. Some depots were converted after steam use to be used by modern traction and have survived to the present time. Some steam sheds had large allocations of certain classes and there were even cases of a shed with only one class of locomotive housed. The steam age has now passed and memories are fading but hopefully the film does not, so here is a picture of BR in the golden age of steam when people and the nation's freight went by rail. A wonderful age of mechanical signalling, antiquated signalboxes, grumpy porters and gas-lit stations.

London Midland Region

WILLESDEN

1A

Sample Allocation 1959
Camden 1B

Class 7P 4-6-0	45514/22/23/32
Class 6P/5F & 7P 4-6-0	45592/45601/06/69/76/86/45722/35
Class 7P 4-6-0	46100/39/44/6/54/61/2/8/70
Class 8P 4-6-2	46229/39/40/2/5/7/54/6
Class 3F 0-6-0T	47302/4/7/10/48/95/47514/22/ 29/47668/9/71

Total 41 Engines

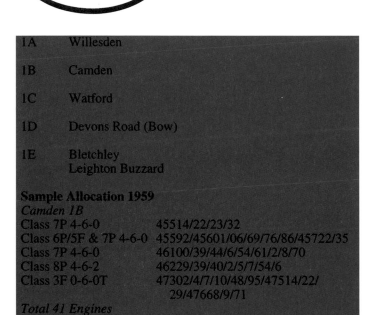

Camden depot (1B) had a good allocation of 'Royal Scot' class 4-6-0s with 17 locos of the class being shedded there in 1950 and nine in 1959. An unidentified 'Royal Scot' heads northwards through Carpenders Park in January 1959 with a West Coast express. *Author*

Sheds in the district originated from the LNWR and provided power for the West Coast main line — Willesden (1A, closed 1965) was the principal freight shed in the London area and had an allocation of 130 engines in 1959 including 27 Stanier '8F' 2-8-0s. A few LNWR 0-8-0s survived in 1959 and a member of the class, No 49395, has survived into preservation. Willesden shed dated from the 1870s but was enlarged and rebuilt under the LMS, which provided a roundhouse in 1929.

Camden (1B, closed 1962) was originally the principal shed and the London & Birmingham Railway roundhouse used in the early days still survives, having been used as a theatre in recent years . Camden LMS shed provided power for the expresses out of Euston. The ill-fated Pacific No 46202 *Princess Anne* was shedded here until wrecked in the 1952 Harrow disaster. Another famous engine was No 46229 *Duchess of Hamilton*, now preserved.

A shed existed at Watford (1C, closed 1965) and this was an LNWR straight shed of six roads, the 1959 allocation being 20 engines.

Devons Road (1D, closed 1964) was the North London Railway establishment which was completely dieselised in 1958. The shed was coded 13B from 1948 to 1949.

Bletchley (1E, closed 1965) had an allocation of 54 engines in 1959 and was coded 2B until 1950, 4A until 1952 and 1E until 1965. Watford and Bletchley had sheds from the earliest days.

Above:
Stanier Pacific No 46229 *Duchess of Hamilton* **was a Camden engine (1B) in 1959 and was one of eight 'Princess Coronation' class 4-6-2s to be allocated to that depot for main line work. The engine has been preserved in BR maroon livery and is the property of the National Railway Museum.** *Author*

Below:
Standard Class 4 2-6-4T No 80084 in clean condition was a Bletchley (1E) engine in 1959 and worked local passenger trains until West Coast main line electrification. The engine was subsequently transferred to the Southern Region. No less than 15 examples of the class have been preserved. *J. Phillips*

2A RUGBY

2A	Rugby
2B	Nuneaton
2D	Coventry
2E	Northampton
2F	Woodford Halse

Sample Allocation 1959
Rugby 2A

Class 4P 4-4-0	41162
Class 2 2-6-2T	41214/78
Class 2P 0-4-4T	41902/9
Class 4 2-6-4T	42061/2/42541/73/77/42669/73
Class 4F 0-6-0	44064/44395
Class 5 4-6-0	44711/15/16/44831/3/6/60/2/3/6/7/70/ 44909/15/38/45146/45493
Class 2 2-6-0	46420/46/72
Class 3F 0-6-0T	47269
Class 8F 2-8-0	48012/8/35/85/48131/6/73/48203/48365/ 48423/7/37/48526/59/48646/68/48757
Class 7F 0-8-0	49245/66/49377/49442
Class 2F 0-6-0	58199/58218/58308

Total 59 Engines

Woodford Halse shed (2F) was the Great Central shed for an important junction on that former system and is seen here on 24 March 1963 with a few BR Standard classes on shed. The London Midland Region took over control of the former GCR line in 1958. *A. N. H. Glover*

Rugby shed (2A, closed 1965) had an allocation of 98 engines in 1950 and 59 in 1959. With main line electrification the shed was recoded 1F from 1963 until closure. The shed was of LNWR origin with 12 roads and was rebuilt by the LMR in 1955. The 1959 allocation shows the last resident compound 4-4-0 No 41162 and two of the Stanier 0-4-4Ts Nos 41902 and 41909.

Nuneaton (2B, closed 1966) had 73 engines in 1950, 62 in 1959 and was partly rebuilt by the LMR that year. The shed had 20 ex-LNWR 0-8-0s in 1959 and received five 'Patriots' in 1960. The shed was coded 2D from 1948 to 1950, 2B from 1950 to 1963 and 5E from 1963 to 1966.

Coventry (2D, closed 1959) had 12 locomotives in 1950. Inherited from the LNWR, the shed's roof was renewed in 1957, some three years before complete closure. Half of its 1950 allocation consisted of ex-LNWR 0-8-0s.

Northampton (2E, closed 1965) had a shed from the earliest days of the London & Birmingham Railway and replaced Wolverton as a running shed. The LNWR shed which housed 40 engines in LNWR days had 37 by 1959. The shedcode changed from 2C to 4B in 1950, 2E from 1952 and 1H in 1963.

Woodford Halse (2F, closed 1965) was of Great Central origin and was coded 38E until 1958 when the LMR took over the GCR line. The code was altered to 2G in 1958 and then 2F until 1963 when it was altered to 1G. Old GCR types and LNER classes gave way to LMR types after 1958. The 1950 allocation was for 54 engines, the 1959 allocation consisted of 51 engines and the final year (1965) consisted of 18 engines. In 1959 26 'WD' class 2-8-0s were allocated.

Top:
A double-header heads north through Lichfield with the 11.50am Euston to Crewe in April 1955. No 41105 was a Rugby (2A) engine and was one of the LMS post-Grouping three-cylinder Compound 4-4-0s. It is seen here with 'Jubilee' class 4-6-0 No 45588 *Kashmir.*
E. S. Russell/Colour-Rail

Above:
A very grimy '8F' 2-8-0 No 48343 is seen arriving at Chester on 4 March 1967. The engine was a Nuneaton (2B) engine in 1959 and a member of Stanier's standard freight type, 666 of which were in BR stock in 1959. *Author*

BESCOT

3 A

3A	Bescot
3B	Bushbury
3C	Walsall
3D	Aston
3E	Monument Lane

Sample Allocation 1959
Bushbury 3B

Class 2 2-6-2T	41225/79
Class 4 2-6-4T	42428
Class 4F 0-6-0	44027/44439
Class 5 4-6-0	44829/45015/45287/45310/45395/ 45405/39
Class 6P & 7P	45555/45647/88/45709/34/37/38/41/2
Class 3F 0-6-0T	47363/97/8/47473
Class 7F 0-8-0	48950/49037/44/49240/49411/52
Class 2F 0-6-0	58118/9/24/83/58204/81/95
Total 38 Engines	

Bescot (3A, closed 1966) had a generous helping of LNWR 'G2' 0-8-0s in 1950 and 39 were allocated at this freight shed at the time. Bescot marshalling yard was one of the largest on the system and the shed had 67 engines in 1950 and 81 by 1959. In 1965 the allocation was down to 62 engines but '8Fs' had replaced the 'G2s'. The shedcode changed from 3A to 21B in 1960 and from 21B to 2F in 1963.

Bushbury (3B, closed 1965) was of LNWR origin and housed 41 engines in 1950 and 38 in 1959. A few of the old LNWR 0-8-0s were still in use in 1959 as well as nine 'Jubilee' class 4-6-0s. The code changed from 3B to 21C in 1960 and from 21C to 2K in 1963.

Walsall (3C), which had an allocation of 57 engines in 1950 was closed in 1958.

Aston (3D, closed 1965) had 52 engines in 1950 and 43 in 1959. LMS types and class 4MT 2-6-0s in the 76000 series were allocated. The shedcode changed from 3D to 21D in 1960 and from 21D to 2J in 1963.

Monument Lane (3E, closed 1962) was a six-road straight shed of the LNWR pattern and housed 32 engines in 1950 and 28 in 1959. The code changed from 3E to 21E in 1960.

'Jubilee' class No 45734 *Meteor* **heads south to Euston through Carpenders Park in January 1959 with the 9.40am Wolverhampton to Euston. Bushbury (3B) had nine 'Jubilee' class 4-6-0s allocated in 1959 for express passenger work.** *Author*

Below:
London & North Western Railway 0-8-0 No 48930 of Bescot (3A) is pictured at Sutton Coldfield with a special working in June 1962. The engine was a member of the LNWR standard heavy freight class known as 'G2A' — a rebuilt 'G1' class of 1912 but with Stanier chimney and a tender cab. *A. N. H. Glover*

Bottom:
Bescot (3A) 0-8-0 No 49361 is seen at Lichfield Trent Valley Junction on 23 June 1963 with an SLS special. The LNWR 0-8-0s in the 'G2A' class were known by railwaymen as 'Super Ds' or 'Duck Eights'. The photographer was left behind after taking the shot. *A. N. H. Glover*

5A CREWE NORTH

Sample Allocation 1959

Crewe North 5A

Class 2P 4-4-0	40652/3/5/9/60/79
Class 2 2-6-2T	41212/20/29
Class 4 2-6-4T	42575/78/42677
Class 6P/5F 2-6-0	42946/54/5/8/61/3/6/8
Class 5 4-6-0	44678/9/80/82–85/44714/58/9/ 60–66/45004/21/33/73/93/45113/ 48/89/45235/40/3/50/4/7/82/9/ 45305/11/48/69/73/9/90/45434/46
Class 6P/5F & 7P 4-6-0	45501/3/28/9/45/6/8
Class 6P/5F & 7P 4-6-0	45553/6/91/45604/23/5/9/30/4/ 43/55/66/74/84/9/45703/21/6/36
Class 7P 4-6-0	46101/10/6/8/20/5/8/9/34/5/8/ 50/1/2/7/9/63
Class 8P 4-6-2	46205/6/12
Class 8P 4-6-2	46220/1/5/8/33/4/5/41/3/6/8/9/ 51/2/3
Class 8P 4-6-2	71000
Class 2 2-6-0	78030
Total 125 Engines	

Crewe North (5A, closed 1965) was one of the principal sheds on the West Coast route and the 1959 allocation shows 125 engines of which 42 were Stanier 'Black 5s'. Stanier Pacifics included three 'Princesses' and 15 'Coronations', of which No 46235 *City of Birmingham* has been preserved. Seven 'Patriots', 19 'Jubilees' and 17 'Royal Scots' were also on the allocation list as well as the solitary BR Class 8 Pacific No 71000 *Duke of Gloucester* which is now up and running again.

A shed had existed at Crewe since Grand Junction Railway days, but the LNWR opened another new shed in 1897 which was known as Crewe South (5B, closed 1967). This was primarily for goods engines, with the allocation in 1950 being for 103 and the allocation in 1959 for 117. LMS freight types predominated but the shed did have two Aspinall L&YR 0-4-0STs, Nos 51204 and 51221, in the 1950 list. The final allocation was for 59 engines, including four Standard Class 2MT 2-6-0s in the 78000 series.

Stafford (5C, closed 1965) was a six-road shed that was completely rebuilt by the LMS in brick and concrete. Some 23 engines were allocated in 1950 and 24 in 1959.

Stoke (5D, closed 1967) was originally the North Staffordshire Railway establishment in that town and consisted of two separate buildings on opposite sides of the running lines. One shed was a roundhouse and the other a straight shed. Stoke accommodated 100 engines in 1950 but these were reduced in number to 71 by 1959.

Alsager (5E, closed 1962) had 18 engines in 1950 and 1959.

Uttoxeter (5F, closed 1964) had six engines in 1950 and seven in 1959. Both 5E and 5F were of North Staffordshire origin.

Stanier Pacific No 46241 *City of Edinburgh* ascends Shap near Scout Green where an LNWR signalbox controlled the crossing. The grubby Pacific in BR green was a Crewe North (5A) engine in 1959. *Author*

Above:
Crewe North (5A) had 15 'Princess Coronation' class 4-6-2s for West Coast main line work in 1959 and a member of the class, No 46225, is seen at Tebay on 31 May 1963 with a stopping train from Carlisle. *Author*

Below:
The *Duke of Gloucester* was a unique engine and the only BR Standard Class 8 Pacific. This was a Crewe North (5A) engine and is seen here at an exhibition at Marylebone in 1961. The engine has been restored and can be seen today at work on special trains. *T. Linfoot*

6A	Chester (Midland)
6B	Mold Junction
6C	Birkenhead
6D	Chester (Northgate)
6E	Chester (West)
6F	Bidston
6G	Llandudno Junction
6H	Bangor
6J	Holyhead
6K	Rhyl

Sample Allocation 1959
Birkenhead 6C

Class 850 0-6-0PT	2012
Class 2021 0-6-0PT	2069
Class 3 2-6-2T	40101/2/21/31/40202/9
Class 2 2-6-2T	41324
Class 4 2-6-4T	42447/93/42597/9/42608
Class 6P/5F 2-6-0	42778/42856/88/94/42941
Class 6P/5F 2-6-0	42969/70/77/78
Class 5 4-6-0	44917
Class 0F 0-4-0ST	47005/9
Class 2F 0-6-0T	47160/4
Class 3F 0-6-0T	47324/38/47431/97/47507/30/65/47627/ 74/7
Class 8F 2-8-0	48120/48260/48349/48448/55/48684/91
Class 5 4-6-0	73032/9
Class 4 2-6-4T	80062/3/90
Class 2 2-6-2T	84000/3
Class WD 2-8-0	90173/90212/90369/92
Total 56 Engines	

Chester LNWR shed (6A, closed 1967) had 38 engines in 1950 and 46 in 1959 of 10 different classes. A rarity on the list was the 0-4-0ST No 47006, one of a class of 10 built by Kitson to a Stanier design.

Mold Junction (6B, closed 1966) had 39 engines in 1950 and 44 in 1959. The final allocation in 1965 consisted of 28 engines of LMS origin.

Birkenhead (6C, closed 1967) was a joint shed between the GWR and the LNWR and was recoded from 8H in 1963. The joint shed was in fact two sheds side by side until Nationalisation in 1948. The LMR took over in 1951 but the variety of classes continued, there being 16 classes out of the 56 engines in use in 1959. In 1950 there were 93 engines including six 'Granges' as well as the last Dean pannier tank, No 2069, which lasted until 1960.

Chester Northgate (6D, closed 1960) was the Cheshire Lines Committee's shed and consisted of a two-road straight building in brick. The LMR takeover ensured that all the old GCR types were replaced and the 1959 allocation of 11 engines included five Class 2 2-6-0s in the 78XXX series.

Chester West (6E, closed 1960) was the ex-GWR shed and was known as 84K until 1958. In the 1950 allocation there were 57 engines which had decreased in number to 55 by 1959. From the LMR takeover in 1958 many of the old GWR types were replaced by LMS or Standard classes. The GWR had two sheds at Chester, one of which was purchased from the LNWR, but both buildings were known as Chester West.

Bidston (6F, closed in 1963) was originally a GCR shed and had 14 engines by 1959.

Llandudno Junction (6G, closed 1966) was coded as 7A until 1952 and was of LNWR origin. A total of 31 engines were shedded there in 1950 and 38 in 1959.

Bangor (6H, closed 1965) was known as 7B until 1952 and had 32 engines in 1950 which decreased to 22 by 1959. The shed was the last haunt of the ex-LNWR 'Cauliflowers' (Nos 58375 and 58381) — a widespread 0-6-0 class of which not a single example was preserved.

Holyhead (6J, closed 1966) was known as 7C until 1952 and had 23 engines in 1950 and 19 in 1959. The shed had five 'Britannia' class Pacifics in 1959 including the unnamed No 70047.

Rhyl (6K, closed 1963), coded as 7D until 1952, had 14 engines of nine classes.

Centre right:
Class 5 No 44917 in extremely grubby condition eases out of Chester General with a Birkenhead train in the last days of steam on the London Midland Region. The engine was allocated to Birkenhead (6C) in 1959 and can be seen under the LNWR signals for which Chester was famous. *Author*

Bottom right:
Mold Junction had an allocation of mainly freight engines including 'Black 5s' and '8Fs'. No 45055 was a Mold engine in 1959 and is seen here double-heading with 'Jubilee' class No 45742 Connaught (the last of class to be built), leaving Penrith with the 2pm Glasgow to Liverpool in 1964. *Author*

Right:
Standard Class 5 4-6-0 No 73026 heads for Birkenhead out of Chester General. The Chester West shed (6E) was the old GWR shed which was known as 84K until handed over to the London Midland Region in 1958. *J. Phillips*

EDGE HILL

8A	Edge Hill
8B	Warrington (Dallam)
	Warrington (Arpley)
8C	Speke Junction
8D	Widnes
8E	Northwich
8F	Springs Branch (Wigan)
8G	Sutton Oak

Sample Allocation 1959

Edge Hill 8A

Class 4 2-6-4T	42121/55/42441/59/42564/70/ 83/42602
Class 5 4-6-0	44768/9/72/3/44906/7/45005/32/39/ 69/45181/45242/9/56/76/45343/76/ 80/98/9/45401/10/3/21
Class 6P/5F & 7P 4-6-0	45515/6/8/21/5/7/31/4/5/9/44/9/50
Class 6P/5F 4-6-0	45552/4/60/67/83/6/96/45670/8/ 81/45733
Class 7P 4-6-0	46114/9/23/4/32/42/7/55/6/64
Class 8P 4-6-2	46200/3/4/7/8/9/11
Class 3F 0-6-0T	47353/57/47402/4/7/11/16/47487/ 8/9/98/47519/66/97/47656
Class 8F 2-8-0	48152/48249/80/48318/48433/ 57/79/48504/9/12/13
Class 7F 0-8-0	49082/49116/32/7/73/49200/49224/ 49355/66/75/92/4/9/49404/5/12/ 16/19/27/9/34/5/7/45

Total 124 Engines

Edge Hill shed (8A, closed 1968) was the principal depot for passenger locomotives in the Liverpool area. The site was in use from the earliest times, having been used by the Liverpool & Manchester and Grand Junction Railways. The GJR had to pay rent to the L&MR and eventually moved to a new shed and workshops at Crewe in 1843. The LNWR enlarged the premises in 1864 and 1902, by which time 118 engines were allocated. The LMS rebuilt the shed and by the time BR was formed there were 19 roads to the shed which held 112 locomotives. In 1959 there were 124 engines including 'Scots', 'Patriots', 'Jubilees' and 'Princesses'. No 46203 *Princess Margaret Rose*, now preserved, was on the 1959 list.

Warrington (8B, closed 1967) shedded mainly ex-LMS freight types with a total of 53 engines in 1959.

Speke Junction (8C, closed 1968), with an allocation of 57 engines in 1965, an increase from the 33 of 1959, had mainly freight types but also had one of the named 'Black 5' class 4-6-0s on the books, No 45154 *Lanarkshire Yeomanry*, as well as '9F' 2-10-0s.

Widnes (8D, closed 1964) had 28 engines in 1959, including two ex-GCR 0-6-0s in the 'J10' class. The shed also possessed the 0-4-0ST No 51218, the L&YR saddletank later preserved on the Keighley & Worth Valley Railway.

Northwich (8E, closed 1968) was recoded from 9G to 8E in 1958, having been 13D until 1950. This old Cheshire Lines shed housed 29 engines in 1959, including GCR types.

Springs Branch (8F, closed 1967) was coded as 10A until 1958 and originated from an 1881 rebuild of an earlier establishment. BR rebuilt the shed in 1950 and by 1959 the shed housed 67 engines, including 32 ex-LNWR 0-8-0s and five ex-GCR 'J10' 0-6-0s.

Sutton Oak (8G, closed 1967) was coded 10E to 1955, 10D to 1958 and finally 8G until closure. The shed had an allocation of 31 engines in 1959 including four LNWR 0-8-0s and a solitary L&YR 0-6-0ST, No 51441.

Lancashire & Yorkshire Railway 0-4-0ST No 51218 of Aspinall's dock shunter class is now resident on the Keighley & Worth Valley Railway. One of the L&YR 'Pugs' to have survived into preservation, the engine was originally L&YR No 68 having been built at Horwich in 1901 and withdrawn by BR in 1964. The engine was allocated to Widnes (8D) in 1959. *Author*

Above:
Sutton Oak (8G) had a clutch of Standard 4MT 2-6-0s in 1959 including No 76077 seen here on a railtour around Widnes in 1967. The docks line is off to the left and the main line is at the rear. The LNWR-type signalbox controlled the square crossing. *C. H. A. Townley*

Below:
No 46201 *Princess Elizabeth* was an Edge Hill (8A) engine on the 1950 allocation list and the second Stanier Pacific to be built in 1933. An O gauge model of the engine was made after the successful record breaking runs in 1936. The engine is one of two members of the class to have been preserved and is seen here on a special excursion near Church Stretton. *Author*

⬭ 9A LONGSIGHT

9A	Longsight (Manchester)
9B	Stockport (Edgeley)
9C	Macclesfield
9D	Buxton
9E	Trafford Park Glazebrook
9F	Heaton Mersey Gowhole
9G	Gorton Ardwick Dinting Guide Bridge Mottram Reddish

Sample Allocation 1959
Longsight 9A

Class 3 2-6-2T	40076/7/8/84/93/40107/22
Class 2P 4-4-0	40674/93
Class 2 2-6-2T	41217/21
Class 2P 0-4-4T	41907/8
Class 4 2-6-4T	42369/81/98/9/42416
Class 6P/5F 2-6-0	42772/86/42814/48/58/87/9/42923/ 4/5/30/4/6/8
Class 4F 0-6-0	44061/44349
Class 5 4-6-0	44686/7/44741/2/6/8/9/50/1/2/ 44827/37/45109/11/50/45302/ 45426
Class 6P/5F & 7P 4-6-0	45505/20/30/6/40/3
Class 6P/5F 4-6-0	45578/87/95/45631/8/44/71/80
Class 7P 4-6-0	46106/8/11/15/22/31/37/40/3/53 46158/60/6/9
Class 3F 0-6-0T	47267/91/47341/3/5/7/56/69/ 95/47400/47528/47673
Class 8F 2-8-0	48165/48275/48389/48428/65/ 48500/48680/48744
Class 7F 0-8-0	49428/39
Class 7P/6F 4-6-2	70031/2/3/43
Total 105 Engines	

Longsight (9A, closed 1965) was the principal LNWR passenger engine shed in Manchester housing 129 engines in 1950 and 105 in 1959. The allocation included 'Patriots', 'Scots', 'Jubilees' and 'Britannias'. No 46115 *Scots Guardsman*, now preserved, was included in the allocation. The shed was in two parts, with the north shed being used for the passenger engines. A shed existed on the site in 1841 and included a roundhouse. Enlargements were made in the 1860s and 1903 by the LNWR. BR rebuilt the south shed in 1957 to accommodate diesels and provided a six-road shed for that purpose.

Stockport Edgeley (9B, closed 1968) had 26 engines in 1959 including 'Crabs' and LNWR 0-8-0s. By 1965 three 'Jubilees' were shedded, including the famous No 45596 *Bahamas*, now preserved.

Macclesfield (9C, closed 1961) was originally a NSR shed with 11 engines in 1950 and six in 1959, all of the same class (Fowler 2-6-4Ts).

Buxton (9D, closed 1968) was recoded to 9L in 1963 and housed 35 engines in 1959. The shed was rebuilt by the LMS in 1935 to replace the Midland Railway building.

Trafford Park (9E, closed in 1968) was of CLC origin and survived until 1968 under BR. The 1950 allocation was for 73 engines and included ex-GCR 'Directors' as well as MR types. The 1959 allocation shows 53 engines, including six 'Britannias'. The code changed from 19F to 13A in 1949, 13A to 9E in 1950, 9E to 17F in 1957 and 17F to 9E in 1958.

Heaton Mersey (9F, closed 1968) was recoded 19D to 13C in 1949, 13C to 9F in 1950, 9F to 17E in 1957 and 17E to 9F in 1958. The 1959 allocation was for 55 engines, mainly ex-LMS '4Fs' and '8Fs'.

Gorton (9G, closed 1965), of GCR origin, was one of the largest on that system and had 166 engines in 1950. By 1959 the number was down to 113 and by 1965, 48. The 1950 allocation included 53 Robinson '04/01' classes which were phased out after the LMR takeover. The shed was recoded 39A to 9H in 1958 and finally 9G in the same year.

Right:
Thompson 'L1' class 2-6-4T No 67756 rolls into Kimberley on the former Great Northern lines in Nottinghamshire. The engine was a Gorton (9G) engine in 1959, a former GCR shed that could boast an allocation of 53 GCR 2-8-0s in 1950. *J. Phillips*

Above:
Class 2 2-6-2T No 41321 was a Gorton (9G) engine in 1959 but is seen here at Seaton in Rutland in 1960 with the 11.10am to Uppingham. The GCR loco classes were replaced by LMR types after 1958. *J. Phillips*

Below:
No 45540 *Sir Robert Turnbull* was an Ivatt rebuild of the Fowler 'Patriot' class known to railwaymen as 'Baby Scots'. No 45540 was a Longsight (9A) engine but is seen here on shed at Saltley (21A) in 1962. *A. N. H. Glover*

⬭ **BARROW**

The oval logo reads:
11
A

11A	Barrow
11B	Workington
11C	Oxenholme
11D	Tebay

Sample Allocation 1959
Tebay 11D
Class 4 2-6-4T	42393/6/42403/4/24
Class 4 2-6-0	43011/28/9/35
Class 4F 0-6-0	44083/44345
Total 11 Engines	

Barrow shed (11A, closed 1966) was the principal shed of the former Furness Railway and was situated next door to the company's works. The code changed from 11B to 11A in 1958, from 11A to 12E in 1960 and from 12E to 12C in 1963. The old Furness types had been disposed of by 1950 when the allocation was 50 engines. By 1958 the engine shed accommodated 48 engines which included nine ex-Midland Railway '2F' 0-6-0s. By 1965 the allocation was down to 21 engines of ex-LMS standard types.

Workington (11B, closed 1968) was of LNWR origin and housed 27 engines in 1950, 31 in 1959 and 22 in 1965. LMS standard goods types predominated and the code was changed several times during the shed's existence. The shed was known as 12D from 1948 to 1955, 12C from 1955 to 1958, 11B from 1958 to 1960, 12F from 1960 to 1963 and 12D from 1963 until closure in 1968.

Oxenholme (11C) was another shed of LNWR origin and closed in 1962. The allocation of the four-road shed was eight engines in 1950 and 1959. In 1959 the engines were Class 4 2-6-4Ts used for banking up Grayrigg Bank and the Windermere branch.

Tebay (11D, closed in 1968) had an allocation of 10 engines in 1950, 11 in 1959 and nine in 1965. The locomotives were used for banking freights over Shap. Tebay was coded as 11E from 1948 to 1950, 11D from 1950 to 1960, 12H from 1960 to 1963 and 12E from 1963 until 1968.

Tebay (11D) was at the foot of the strenuous climb to Shap on the West Coast main line. Class 4 2-6-4T No 42095 banks a freight out of Tebay station headed by Class 5 4-6-0 No 44769 on a fine sunny morning in September 1964.
A. N. H. Glover

Above:
Scout Green (half-way up the climb to Shap) with Class 4 2-6-4T No 42225 descending light engine to Tebay (11D) after banking a northbound freight. The LNWR box and crossing gates complete the picture. Note the detonator clamp on the rail in the foreground.
A. N. H. Glover

Right:
Workington (11B) in May 1960 with LNWR bracket signal post and '4F' No 44549 approaching. The '4F' 0-6-0s were one of the LMS's most numerous classes and 580 locomotives were built from 1924 onwards. The design followed the Fowler MR 0-6-0 type of 1911. *Author*

CARLISLE KINGMOOR

12A	Carlisle (Kingmoor)
12B	Carlisle (Upperby) Penrith
12C	Carlisle (Canal)
12D	Kirkby Stephen

Sample Allocation 1959
Carlisle Upperby 12B

Class 2P 4-4-0	40628/9/56
Class 4 2-6-4T	42426/49/42539/94/42664
Class 4F 0-6-0	43896/44016/60/81/44121/6/ 44326/46/44596
Class 5 4-6-0	44770/44936/9/45025/70/45106/ 12/40/85/97/45244/6/8/58/9/ 86/93/5/6/7/45315/16/17/23/ 9/44/51/68/71/94/7/45402/12/ 14/31/37/8/45445/51/94
Class 6P/5F & 7P 4-6-0	45502/7/8/12/13/24/6/33/7/41/51
Class 6P/5F 4-6-0	45588/93/99/45617/72/45723
Class 7P 4-6-0	46126/36/41/65/7
Class 8P 4-6-2	46226/36/7/8/44/50/5/7
Class 2 2-6-0	46449/57
Class 3F 0-6-0T	47288/92/5/47326/37/40/77/ 47408/15/92/47602/14/66
Class 2F 0-6-0	58215

Total 103 Engines

Kingmoor shed (12A, closed 1968) was of Caledonian origin and in 1950 housed 142 engines. In 1959 there were 143 engines and by 1965, 119. The shed was coded as 12A from 1948 to 1949, 68A from 1949 to 1958 and 12A from 1958 to 1968. A good variety of locomotive types were used and the 1959 list is of interest. There were 26 'Crab' class 2-6-0s, 18 'Jubilees', five 'Clans' and no less than 50 of the ubiquitous Class 5 4-6-0s, the workhorses of the LMR system. By 1965 there were 119 engines allocated and the number of 'Black 5' 4-6-0s had gone up to 62 with 16 'Britannia' Class 7 4-6-2s in addition.

Upperby shed was the ex-LNWR establishment and originated from the early LNWR days although rebuilt and modernised by the LMS. The shed at Upperby (12B, closed 1966) included a 32-road roundhouse and had an allocation of 103 engines in 1959, including 11 'Patriots', six 'Jubilees' (including No 45593 *Kolhapur*), five 'Royal Scots' and eight 'Coronations'. The code for the shed was 12B from 1948 to 1950, 12A from 1950 to 1958 and 12B again from 1958 until 1966. The final allocation for 1965 was for 21 engines which included seven 'Britannia' class Pacifics. The Penrith sub-shed was a two-road affair and closed completely in 1962.

The former North British establishment at Canal was known as 12B from 1950 to 1951, 68E from 1951 to 1958, 12D in 1958 and finally 12C from 1958 until 1963, when it closed. In 1959 the shed had 41 engines, which was a reduction on the 59 of 1950. The LNER types predominated with four 'A3' 4-6-2s and 'D49' 4-4-0s being represented. There was a covered roundhouse and straight shed.

The superb stone-built four-road shed at Kirkby Stephen (12D, closed 1961) with wooden doors kept permanently in the open position was of North Eastern Railway origin and a modeller's dream. The shed was coded as 51H from 1949 to 1958, 12E in 1958 and 12D from 1958 to 1961. The shed was transferred to the London Midland Region in 1958 and had an allocation of 11 engines, including 'J21' and 'J25' 0-6-0s, Class 4 2-6-0s, Class 2 2-6-0s (78000) and a solitary Class 2 2-6-0, No 46470.

'Jubilee' class 4-6-0 No 45588 *Kashmir* was a Carlisle Upperby (12B) engine and is seen at Lockerbie in 1963 on a railtour. Carlisle Upperby shed was of LNWR origin and included a 32-road roundhouse which was closed in 1966. *Author*

Above:
The 'Clan' class Pacifics were a rare class of 10 engines allocated between Polmadie (66A) and Carlisle Kingmoor (12A). No 72008 *Clan Macleod*, a Carlisle engine, is seen leaving Perth in June 1963 with an evening Glasgow train. *Author*

Below:
Carlisle Canal (12C) was the North British shed and had an allocation of 'A3' class Pacifics for the Waverley route to Edinburgh. No 60093 *Coronach* is seen waiting to depart northwards in May 1960 from Carlisle Citadel. *Author*

14A CRICKLEWOOD

14A	Cricklewood
14B	Kentish Town
14C	St Albans
14D	Neasden Aylesbury Chesham Marylebone Rickmansworth
14E	Bedford

Sample Allocation 1959
Kentish Town 14B

Class 3 2-6-2T	40021/7–36/8/40/92/40100/11/9/42/ 60/72
Class 2P 4-4-0	40548/67/80/2
Class 4 2-6-4T	42156/78/42237/42325/9/34/8/42/ 42540/87/42610/7/82/5
Class 4F 0-6-0	43964/44052/44210/35/43/70/94/98/ 44381/44531/2/63
Class 5 4-6-0	44658/63/44810/2/7/21/2/5/44846/55/ 44985/45277/9/85/45407/47
Class 6P/5F 4-6-0	45557/61/75/9/85/98/45612/4/5/6/8/ 22/8/49/52
Class 7P 4-6-0	46103/30/3/48
Class 3F 0-6-0T	47200/2/4/5/9/12/29/41/60/83/47437/ 47642/4/5
Class 2F 0-6-0	58131

Total 100 Engines

The Midland's London freight shed was at Cricklewood (14A, closed 1964) and in 1950 it had an allocation of 89 locomotives of Midland and ex-LMS types. By 1959 the allocation had dropped to 54 engines for the two enclosed roundhouses.

Kentish Town (14B) was the Midland's passenger shed for the London area, and this closed in 1963. The shed consisted of three enclosed roundhouses and accommodated 117 engines in 1950 and 100 in 1959. Notable named classes were the 15 'Jubilees' and the four 'Royal Scots'. Other ex-LMS types included 19 Fowler Class 3 2-6-2Ts used on the St Pancras suburban services. These engines were fitted with condensing equipment to work through to Moorgate.

St Albans shed (14C, closed 1960) had an allocation of 17 engines of ex-LMS types and was a two-road straight brick-built affair with wooden end doors.

The LMR took over the GCR lines in 1958 and Neasden shed became coded 14D from 34E until closure in 1962. The shed at Neasden always had a cosmopolitan collection of engines and the 1950 allocation of 82 locos included GCR and GWR types. The 1959 allocation of 70 engines included BR Standard types and Class 4 2-6-4Ts which replaced the LNER 'L1' 2-6-4Ts. Neasden could accommodate visiting engines and on occasions representatives from five of the BR Regions could be seen.

Bedford (14E, closed 1963) was a Midland four-road shed with modifications by the LMS and BR. The shed was coded 15D from 1948 to 1958, 14E from 1958 to 1963 and 14C in the final year of 1963. The 1950 allocation was of 42 and the 1959 allocation was for 31 engines. Old Midland types were replaced by BR Standards in the 1960s.

Right:
Bedfordshire branch lines were worked by old Midland 0-4-4Ts that were replaced by Ivatt 2-6-2Ts and Standard Class 2 2-6-2Ts. No 84005 (14E) is seen entering Shefford with a Hitchin train in 1961 shortly before closure of the Bedford to Hitchin line.
J. Phillips

Left:
The Midland Railway had a large class of 0-6-0s originating from the Johnson design of 1885, rebuilt by Fowler with Belpaire firebox and classified by the LMS as '3F'. No 43474 is seen on Bedford (14E) shed in March 1961. *Author*

Below:
Nottingham Victoria with Class 5 4-6-0 No 44847 of Neasden (14D) about to depart on the 8.15am to Marylebone on 6 August 1966 shortly before the GCR mai line closed. The LMR took over the GCR in 1958 and substituted LMR types of motive power. *A. N. H. Glover*

⬭ 15A WELLINGBOROUGH

Sample Allocation 1959

Leicester Midland 15C

Class 2P 4-4-0	40402/52/40543
Class 4 2-6-4T	42137/60/82/42330/31
Class 3F 0-6-0	43205/61/77/43326/43405/11/ 43629/43728/99
Class 4F 0-6-0	43919/37/44034/44231/44403/23
Class 5 4-6-0	44667/90/44811/15/43/48/45264
Class 2 2-6-0	46454
Class 3F 0-6-0T	47213/50/74/47313/47441/2/47533/4/43
Class 8F 2-8-0	48007/10/27/61/48107/33/49/48211/66
Class 4 2-6-0	75057/58/59/60/61
Class 2 2-6-0	78029
Class 9F 2-10-0	92100–4/9/21/28

Total 63 Engines

Wellingborough (15A, closed 1966) was a freight shed and of Midland Railway origin. The shed was coded 15A from 1948 to 1963 and 15B from 1963 to 1966. The 1950 allocation of 77 engines consisted of LMS standard types including 47 Stanier '8F' 2-8-0s. A few Midland relics survived including three '1P' class 0-4-4Ts dating from the Johnson era. The 1959 allocation of 63 engines included 35 '9F' class 2-10-0s of which 10 were the Franco-Crosti-boilered version, a most unusual looking design of Italian origin. The shed consisted of two enclosed roundhouses, one of which was demolished in 1964 to provide a new diesel depot.

Kettering (15B, closed 1965) was a four-road brick-built shed of Midland origin. The 1950 allocation of 37 engines was of standard LMS types and the 1959 allocation of 39 engines included 19 Stanier '8F' 2-8-0s as well as five Class 9 2-10-0s and three Class 2 2-6-0s (78000 series) for working the line to Cambridge.

Leicester Midland shed had 80 engines in 1950 and 63 in 1959; a noticeable feature of the shed was the 32-road roundhouse. The shed was rebuilt from 1952 and a new concrete coaling stage installed. Leicester Midland (15C, closed 1966) was recoded in 1963 to 15A. A few Midland types were in existence in 1959, mainly '3F' class 0-6-0s.

Coalville (15D, closed 1965) was known as 17C from 1948 to 1958, 15D from 1958 to 1963 and 15E from 1963 to 1965. Thirty-three engines were allocated in 1950 and 24 in 1959. The three-road Midland shed housed mainly '3F' and '4F' 0-6-0s but the 1965 allocation shows 17 '8Fs' as the entire stock in use.

Leicester Central (15E, closed 1964) was the Great Central shed recoded from 38C in 1958 after the LMR takeover to 15E and altered to 15D in 1963. The shed had 23 engines allocated in 1959 and was a four-road straight building. The allocation in 1950 included eight 'A3' class Pacifics, including the renowned No 60103 *Flying Scotsman*. By 1959 'B1s' had replaced the 'A3s' but the allocation included two named examples: No 61008 *Kudu* and 61028 *Umseke*.

Market Harborough (15F, closed 1965) was a delightful set-up as it was a two-road shed of LNWR origin with the water tank forming the roof. Four engines were allocated in 1959 and the code changed from 2F to 15F in 1958.

Uppingham church is seen in the background as Class 2 2-6-2T No 84006 of Wellingborough (15A) passes through the Rutland countryside *en route* to Seaton. The 2-6-2Ts were of a small class of 30 engines, none of which were preserved. *J. Phillips*

Seaton, Rutland was a country junction for the Uppingham branch which was built by the LNWR to rival the nearby Midland main line. No 84006 (15B), a BR Standard Class 2 2-6-2T, waits to take the branch train on 26 September 1959. *Author*

Below:
Class 9F No 92154 stops to shunt at Kings Heath in 1966, a Midland Railway station that closed in 1946. The Class 9F 2-10-0 was shedded at Wellingborough (15A) in 1959 and was one of the 35 examples of that class allocated to that shed, including the Franco-Crosti-boilered types. *A. N. H. Glover*

NOTTINGHAM

16A	Nottingham
16B	Kirkby-in-Ashfield
16C	Mansfield
16D	Annesley
	Nottingham (Victoria)
	Kirkby Bentinck

Sample Allocation 1959
Peterborough Spital Bridge 31F (16B until 1950)
Class 3P 4-4-2T	41949/69/75
Class 3 2-6-0	43127
Class 4F 0-6-0	43957/44097/44110/52/44239/47/73/76/ 44509/18/19/21/22
Class 3F 0-6-0T	47300
Class B1 4-6-0	61095/6/61156/61204/5/61323/48
Class D16 4-4-0	62597/62612/3
Class J39 0-6-0	64789/64901
Class WD 2-10-0	90063/90447/90501/28
Total 34 Engines	

Nottingham shed (16A, closed 1965) consisted of three roundhouses originating from the Midland Railway, 144 engines being allocated in 1950 and 96 in 1959. The code changed from 16A to 16D in 1963. The shed housed six 'Jubilees' and 13 '2P' class 4-4-0s in 1959 with 38 '4F' class 0-6-0s of MR and LMS origin. There were also 21 '8F' class 2-8-0s allocated in the same year.

Kirkby-in-Ashfield (16B, closed 1966) had 63 engines in 1950 of which 42 were '8F' 2-8-0s. By 1959 the allocation was down to 48 engines, of which 13 were '4F' 0-6-0s and 34 were '8F' 2-8-0s with a solitary '3F' 0-6-0. The code was changed from 16C to 16B in 1955 and from 16B to 16E in 1963. The shed was enlarged to a five-road building in 1958.

Mansfield (16C, closed 1960) was a four-road straight brick-built Midland shed with 28 engines in 1950 and 29 in 1959. The 1959 allocation included 15 '8F' 2-8-0s used for coal trains in the area. The code was changed from 16D to 16C in 1955.

Annesley (16D, closed 1966) was the Great Central shed in the area and had an allocation of 77 engines in 1950 and 70 in 1959. The GCR 2-8-0s of the 'O4' and 'O1' classes were very prevalent in 1950 and 53 were allocated. The '9Fs' came in to replace the aged GCR types and eventually the shed ended up with 19 '5MT' 4-6-0s, 13 '8F' 2-8-0s and 29 '9F' 2-10-0s in 1965. The shedcode was changed from 38B to 16D in 1958 and from 16D to 16B in 1963.

'Crab' class 2-6-0 No 42769 was one of the five 2-6-0s to be allocated to Annesley (16D) shed in 1959. The shed was of GCR origin and was recoded to 16B in 1963. The engine was one of the 245 examples in the class at the time and is seen at Warrington Dallam in 1963. *Hugh Ballantyne*

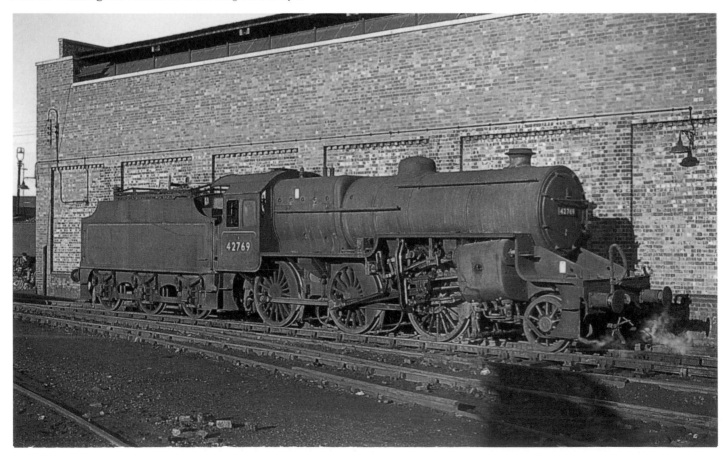

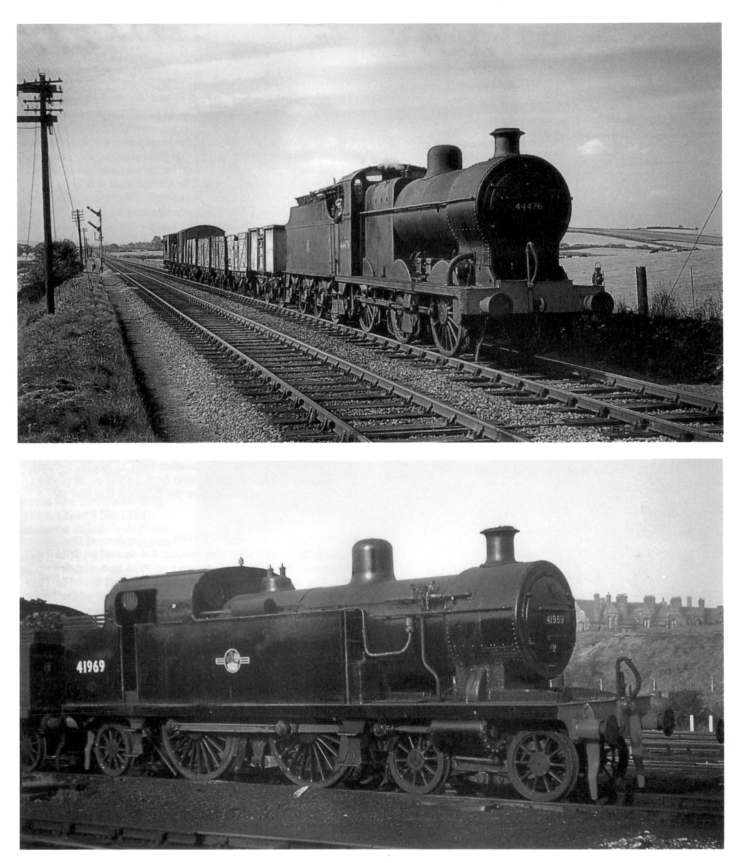

Top:
**'3F' class 0-6-0 No 44476, an engine from a class of 580,
arrives at Seaton in Rutland with an unfitted freight in
September 1959. A long-forgotten sight, the pick-up goods,
would lumber along at very slow speeds. The engine was
shedded at Peterborough Spital Bridge, an old Midland
shed taken over by the Eastern Region in 1950 and recoded
from 16B to 35C.** *Author*

Above:
**The London, Tilbury & Southend Railway 4-4-2T designs
were superseded by the more modern Standard Class 4
2-6-4Ts during the 1950s. As a result of their replacement
the engines were dispersed over the rest of the LMR.
No 41969 was sent from Plaistow (33A) to Peterborough
Spital Bridge (31F) but the class was not very popular there
and was little used. The engine is seen in ex-works condition
at Plaistow in 1958.** *E. V. Fry/Colour-Rail*

⑰A DERBY

Sample Allocation 1959
Derby 17A

Class 2P 4-4-0	40412/16/40513/36/38/40682
Class 4P 4-4-0	40925/41062/41157
Class 1F 0-6-0T	41726/54/73/41847
Class 4 2-6-4T	42174/81/84/42326
Class 3F 0-6-0	43200/43306/15/18/24/68/43459/ 43510/48/84/43658/43727/35
Class 4F 0-6-0	43840/81/43925/30/55/69/91/ 44020/31/42/48/49/44112/42/64/69/ 76/77/44214/95/44304/34/69/80/ 44409/19/20/25/28/65/44540/45/44601
Class 5 4-6-0	44688/44818/39/51/56
Class 6P/5F 4-6-0	45610/26/27/48/63
Class 2 2-6-0	46402/40/43/97/99/46500/02
Class 0F 0-4-0ST	47000
Class 3F 0-6-0T	47563/47629/60
Class 8F 2-8-0	48005/79/83/48121/53/68/48270/93/ 48302/90/48403/48510
Class 2F 0-6-0	58132/44/58/58219
Class 5 4-6-0	73135–44/57–9

Total 113 Engines

Derby (17A, closed 1967) was the headquarters of the Midland Railway, Britain's third largest pre-Grouping railway company, and had the famous works as well as the administrative centre of that vast sprawling system. The engine shed was a commodious enclosed roundhouse with 138 engines allocated in 1950 and 113 in 1959. The code changed to 16C in 1963 and the final year showed 42 engines, mainly '8F' class 2-8-0s. The 1959 allocation included six '2P' class 4-4-0s and three '4P' compound 4-4-0s. Five 'Jubilee' class 4-6-0s were shedded as well as one of the diminutive 0-4-0ST classes, No 47000, a Kitson design of 1932.

Burton (17B, closed 1966), a traditional straight ex-Midland shed, had 108 engines in 1950, 89 in 1959 and 34 in 1965. A few Midland relics pottered around in 1959, the most notable survivor being Johnson '1F' 0-6-0T No 41878. The majority of the locomotives employed in 1959 were '3F' and '4F' 0-6-0s.

Rowsley (17D, closed 1965) was a traditional MR four-road shed with an allocation of 51 engines in 1950 and 53 in 1959. The shedcode changed from 17D to 17C in 1958 and from 17C to 16J in 1963. Standard MR and LMS types could be seen on shed, the most famous locomotives were, however, the ex-North London 0-6-0Ts, of which No 58850 was the last survivor. No 58850 was withdrawn in September 1960. The engine was built at Bow in December 1881 as No 76, became LNWR No 2650 in 1909 and came from a class of 30 engines. The engine can be seen today on the Bluebell Railway in Sussex.

The Midland compound No 1000 was the first of the Johnson three-cylinder Compound 4-4-0s and built in 1902. The engine was rebuilt by Fowler in 1914 and withdrawn in 1951. The engine was restored to the MR red livery in 1959 and worked railtours. The engine was shedded at Derby (17A) during its final years and can be seen here with 'Jubilee' class 4-6-0 No 5690 *Leander* at Sellafield. *Author*

Right:
Ivatt Class 2 2-6-2T No 41277 (17B) is seen at Tutbury in 1959 with the 'Tutbury Jenny', the push and pull train that worked from Burton upon Trent to Tutbury until 13 June 1960. *Author*

Below:
BR Standard Class 5 4-6-0s were allocated to Derby (17A) in 1959 but were displaced by the Midland main line dieselisation in 1961. Class 5 No 73141 makes a steamy exit from Chester in 1967 during the last days of BR steam. *Author*

18A TOTON

Sample Allocation 1959

Toton 18A

Class 3P 4-4-2T	41947
Class 3F 0-6-0	43251/43309/43453/99/43650/43793/43826/31
Class 4F 0-6-0	43845/60/65/43921/90/4/44012/44106/40/61/78/44200/24/84/44376/44427
Class 3F 0-6-0T	47223/47/47551
Class 8F 2-8-0	48099/48118/28/45/83/84/85/86/87/94/5/6/7/48201/21/71/84/48304/6/14/19/24/32/3/8/50/61/2/3/7/70/84/7/90/48507/17/30/38/45/48604/06/07/15/16/20/36/7/40/62/72/81/5/94/98/48728
Class 2F 0-6-0	58153/66/73
Class 9F 2-10-0	92050/57/77/78/86/94/92129/30/31/53/56/58

Total 98 Engines

Toton (18A, closed 1965) was one of the principal freight sheds on the Midland and served the marshalling yard with coal traffic from the Nottingham coalfield. The code of the shed was changed in 1963 from 18A to 16A. LMR freight types were used and the 1950 allocation of 155 in 1950 included 23 Beyer Garratt 2-6-6-2s as well as 57 Class 8F 2-8-0s. By 1959 the Garratts had gone, but the shed housed 55 '8F' 2-8-0s out of the total allocation of 98 engines. Twelve '9F' class 2-10-0s were also allocated as well as 24 0-6-0s of '3F' and '4F' classes. The shed at Toton, which consisted of roundhouses, was repaired and rebuilt by the LMS and BR. Today it is a diesel depot.

Westhouses (18B, closed 1966) was recoded from 18B to 18G in 1963 and had 61 engines in 1950 and 33 in 1959. The shed did not have turning facilities as locomotives were turned on the nearby Blackwell triangle.

The Midland shed at Hasland (18C, closed 1964) had 49 engines in 1950 and 39 in 1959. The code was changed from 18C to 16H in 1963. Ten Beyer Garratts were allocated until their withdrawal in 1957. The shed had a partially covered roundhouse.

Staveley Midland (18D, closed 1965) was an ex-Midland roundhouse. It was coded 18D from 1948 until 1958 and then 41E until closure. Amongst its allocation of small tank engines were a number employed in the adjacent steelworks.

The LMS 0-6-0 shunting tank was a development of the Midland 0-6-0T of 1899 and was produced in large numbers by the LMS. The class totalled 417 engines in 1959 and several examples have been preserved. No 47218 of 18C is seen at Toton (18A) in 1961. *A. N. H. Glover*

Above:
Toton (18A) had a large allocation of Stanier '8F' 2-8-0s to work freight trains around the Midlands and had 57 members of the class allocated in 1950. No 48685 is seen here at Luffenham on an iron-ore train in September 1959. *Author*

Below:
An 1880-vintage Midland 0-6-0T No 41708 has been preserved and is seen at Parkend on an unfitted freight in rural surroundings. It was renumbered 41748 for the occassion. The class consisted of 280 engines built by S. W. Johnson from 1874 to 1900. When the engine was withdrawn in 1966 it was the oldest engine on BR and was for much of its career allocated to Staveley (18D) where engines were hired out to the steel works. *Author*

SALTLEY

21A	Saltley	
21B	Bournville	

Sample Allocation 1959
Saltley 21A

Class 3 2-6-2T	40012/40115/49	
Class 2P 4-4-0	40443/40511	
Class 4 2-6-4T	42054/42327/37/40/83	
Class 6P/5F 2-6-0	42758/61/75/88/90/1/42813/16/23/27/	
	46/57/90/42900/03	
Class 4 2-6-0	43010/13/17/36/41/46/7/9	
Class 3F 0-6-0	43210/14/19/23/42/53/84/43339/55/74/	
	79/81/43433/35/68/82/4/90/43507/94/	
	99/43620/7/44/73/4/80/93/43812	
Class 4F 0-6-0	43858/78/43911/32/8/9/40/9/51/63/5/	
	85/6/44004/13/26/84/91/2/44108/37/8/	
	43/60/65/71/9/84/5/7/44201/3/11/13/	
	26/7/30/63/44333/44406/13/63/44515/	
	20/80/3	
Class 5 4-6-0	44659/60/4/6/44775/6/44804/5/13/4/41/	
	2/59/88/44919/20/45/62/3/4/5/6/	
	45040/45186/45265/8/9/72/80/45333	
Class 8F 2-8-0	48002/48101/5/48220/48315/36/9/42/	
	51/88/48523/48647/69/87/48700	
Class 2F 0-6-0	58168/58261	
Class 9F 2-10-0	92008/9/48/49/51/3/92120/35/6/7/8/9/	
	50/1/2/5/7/65/7	

Total 174 Engines

Saltley (21A, closed 1967) was the principal Midland shed in the Birmingham area and consisted of enclosed roundhouses, rebuilt and repaired by the LMS and LMR. Mainly freight locomotives were allocated and the 1950 figures show 180 engines which had declined in number to 174 in 1959 and 49 by 1965. In 1959 there were 15 'Crab' class 2-6-0s, 29 '3F' 0-6-0s, 46 '4F' 0-6-0s and 19 '9F' 2-10-0s. The shedcode was altered from 21A to 2E in 1963.

Bournville (21B, closed 1960) had 31 engines in 1950 and 22 in 1959. Bournville, an ex-Midland shed, was an enclosed roundhouse and contained a few Midland relics at closure, notably the '2F' 0-6-0s of Johnson design.

Class 2P 4-4-0 No 40439 has seen better days in this scene at Toton in June 1961. The former station pilot at New Street was a Bournville (21B) engine in 1959. None of the LMS 2P 4-4-0s, of which there were 134 in 1959, have been preserved. *A. N. H. Glover*

Above:
Class 5MT 4-6-0 No 44859 leaves Penrith on 1 August 1964 with a Kinross-Liverpool 'Glasgow Fair' holiday special. The engine was allocated to Saltley (21A) in 1959, one of 30 'Black 5s' to be shedded there in the 1950s. *Author*

Below:
The well-known LMS 2-6-4T design had variations and the engine illustrated is one of the Fowler designs dating from 1927. No 42383 is seen in store at Toton in 1961. The locomotive was allocated to Saltley (21A) in 1959.
A. N. H. Glover

⟨24A⟩ ACCRINGTON

24A	Accrington
24B	Rose Grove
24C	Lostock Hall
24D	Lower Darwen
24E	Blackpool
24F	Fleetwood
24G	Skipton
24H	Hellifield
24J	Lancaster (Green Ayre)
24K	Preston
24L	Carnforth

Sample Allocation 1959
Lostock Hall 24C

Class 3 2-6-2T	40192
Class 4 2-6-4T	42158/42286/96/8/42434/76/81/42634
Class 3F 0-6-0	52182/52290/52429/45/56/8
Class WD 2-8-0	90258/66/77/95/90331/5/67/98/
	90413/90541/56/90658/75/81/9/90720

Total 31 Engines

Accrington (24A, closed 1961) was an ex-Lancashire & Yorkshire Railway straight shed of six roads. A shed existed there in 1848 and was erected by the East Lancashire Railway, a constituent of the L&YR. The LMS rebuilt the L&YR shed and the six-road structure was converted into a DMU depot by BR in 1961.

Rose Grove (24B, closed 1968) was one of the last depots in use for steam on BR and survived until August 1968. The 1950 allocation was for 49 engines, with the 1959 total being of 46 engines. The predominant class of engine was the 'WD' 2-8-0, of which 16 were allocated in 1959. The shedcode changed from 24B to 10F in 1963.

Lostock Hall (24C, closed 1968) was another of the last three steam depots in use on BR in August 1968 and had 44 engines in 1950 which became reduced to 31 in 1959. 'WD' class

2-8-0s were in the majority in 1959 with an allocation of 16 engines. The code changed from 24C to 10D in 1963.

Lower Darwen (24D, closed 1966) was another shed originating from the L&YR and had 38 engines in 1950 and 26 in 1959. The code changed from 24D to 10H in 1963. Old L&YR types had been replaced by BR Standard engines by 1959.

Blackpool (24E, closed 1964) was of Lancashire & Yorkshire origin, but was rebuilt by BR in 1958. The eight-road straight shed had an allocation of 61 engines in 1950 and 39 in 1959. The 1959 allocation shows six 'Jubilee' class 4-6-0s and 20 'Black 5' 4-6-0s. The shedcode changed from 24E to 28A in 1950, from 28A to 24E in 1952 and finally from 24E to 10B in 1963.

Fleetwood (24F, closed 1966) had 33 engines in 1950 and 23 in 1959, the old L&YR types having been replaced by ex-LMS and Standard types by 1965. The code changed from 24F to 28B in 1950 back to 24F in 1952 and finally from 24F to 10C in 1963.

Skipton (24G, closed 1967), an ex-Midland shed rebuilt by BR, was a very cramped six-road affair. A total of 36 engines were shedded in 1950 and 24 in 1959. The '4F' class 0-6-0s were the predominant type with 15 of the class being shedded in 1959. The shedcode changed from 20F to 23A in 1950, back to 20F in 1951, to 24G in 1957 and 10G in 1963.

Hellifield (24H, closed 1963) had 23 locomotives in 1950 and 15 in 1959. The Midland four-road shed was used for storage of historic locomotives after closure. The code changed from 20G to 23B in 1950, back to 20G in 1951 and from 20G to 24H in 1957.

Lancaster Green Ayre (24J, closed 1966) was an ex-Midland shed with an allocation of 40 engines in 1950 and 35 in 1959. The code of the shed changed four times during BR days, namely 20H to 23C in 1950, 23C to 11E in 1951, 11E to 24J in 1957 and 24J to 10J in 1963. The shed was home to four of the Stanier '2P' 0-4-4Ts in 1950, a class of only 10 engines.

Preston (24K, closed 1961) was an ex-LNWR shed with 36 engines in 1950 and 26 in 1959. 'Patriots' and 'Jubilees' were allocated as well as the LNWR '7F' 0-8-0s. The shed was noted for the disastrous fire of June 1961 which destroyed the roof and some of the locomotives under it. The code changed from 10B to 24K in 1958.

Carnforth (24L, closed 1968) had an allocation of 42 engines in 1950 and 1959 of mainly ex-LMS types. The shed still exists as the Steamtown Railway Museum crammed full of preserved goodies. The shedcode was changed from 11A to 24L in 1958 and 24L to 10A in 1963.

Centre right:
On the final approach to Shap summit on 1 August 1964 'Black 5' class 4-6-0 No 44730 works the 11.20am Blackpool to Glasgow special unassisted up the steep climb. The engine was allocated to Blackpool (24E) in 1959. *Author*

Bottom right:
The last gasp of main line steam on BR occurred in 1968 and many special trains were run for railway clubs. Nos 44874 and 45017, both Carnforth (24L) engines, work a special on the climb to Copy Pit on 4 August 1968. *Author*

Above:
The '4F' class 0-6-0 was the LMS workhorse for shunting and medium-range freight work. A development of the Fowler Midland design, it could be found all over the system. No 44041, a Skipton (24G) engine, can be seen at Kingmoor in August 1964.
A. N. H. Glover

NEWTON HEATH

26A	Newton Heath
26B	Agecroft
26C	Bolton
26D	Bury
26E	Lees (Oldham)
26F	Patricroft

Sample Allocation 1959
Lees (Oldham) 26E

Class 2 2-6-2T	41206
Class 4 2-6-4T	42114/5/42551/42657
Class 3F 0-6-0	52183/52240/8/69/52322/52410/66
Class 2 2-6-2T	84013
Class WD 2-8-0	90123/40/1/90306/90402/90525/ 90671/90708

Total 21 Engines

Morecambe Promenade in May 1960 sees 'Patriot' No 45509 *The Derbyshire Yeomanry* built at Crewe in 1932 on a special for Manchester. The engines were officially rebuilds from the LNWR 'Claughton' class 4-6-0s but few parts were from the original engines. The engine was shedded at Newton Heath (26A) in 1959 and was the only 'Patriot' allocated there. *Author*

Newton Heath (26A, closed 1968) was the Lancashire & Yorkshire Railway's largest shed with 24 roads. The straight-roaded engine shed was rebuilt by the LMS and in 1950 had an allocation of 167 engines. In 1959 the allocation was down to 154 engines, which included eight 'Jubilees', one 'Patriot' and 35 'Black 5s', of which two were named. The code of the shed changed from 26A to 9D in 1963.

Agecroft (26B, closed 1966) was another ex-L&YR shed which was half demolished by BR in 1954 leaving only four roads under cover. The shed had five of the Fowler 0-8-0s in 1959 as well as two L&YR 0-4-0 and 0-6-0 saddletanks, an example of both classes having been preserved. The 1950 allocation was for 56 engines and the 1959 was for 53. The shedcode changed to '9J' in 1963.

Bolton (26C, closed 1968) housed quite a few old L&YR crocks in 1950 including three Aspinall 0-6-0STs and nine 0-6-0s. The shedcode changed to '9K' in 1963.

Bury (26D, closed 1965) had 28 engines in 1950 and 31 in 1959. The 1950 list included six Fowler '7F' 0-8-0s and five Aspinall '2F' 0-6-0STs as well as 11 '3F' Aspinall 0-6-0s. The shedcode changed from 26D to 9M in 1963.

Lees (26E, closed 1964) was a traditional six-road shed rebuilt by BR in 1955 with an allocation of 23 engines in 1950 and 21 in 1959. Lees was the last haunt of the L&YR '3F' 0-6-0s of 1889 (L&YR Class 27). One member of the class, No 52322, has survived into preservation. The shedcode was changed from 26F to 26E in 1955 and 26E to 9P in 1963.

Patricroft (26F, closed 1968) had an allocation of 73 engines in 1950 and 78 in 1959. The shed had an unusual layout and consisted of two buildings set at different angles to one another. The 1959 allocation included 17 ex-LNWR 0-8-0s as well as six 'Jubilee' class 4-6-0s. The shedcode changed from 10C to 26F in 1958 and from 26F to 9H in 1963.

**Tottington branch sidings with a '3MT' class
2-6-2T, No 40063, waiting for the signal to clear in
July 1962. The engine was one of six '3MT' 2-6-2Ts
to be allocated to Newton Heath (26A) at the
time.** *Author*

**Lancashire & Yorkshire Railway 0-6-0 No 52523
of Bolton (26C) shed is seen at Tottington Sidings
on the former Holcombe Brook branch on a
railtour in 1962. One member of the Aspinall 0-6-0
class, No 52322, has been preserved.** *Author*

BANK HALL

27A	Bank Hall
27B	Aintree
27C	Southport
27D	Wigan
27E	Walton-on-the-Hill
27F	Brunswick (Liverpool)
	Warrington (Central)

Sample Allocation 1959

Southport 27C

Class 3 2-6-2T	40090/40191/4–99
Class 4 2-6-4T	42290/2/3/42435/7/42621/37
Class 5 4-6-0	44728/9/44887/89/45061/45218/28
Class 2P 2-4-2T	50746/81
Class 4 4-6-0	75015–19
Total 29 Engines	

'Black 5' No 44887 was shedded at Southport (27C) in 1959 and can be seen crossing the viaduct at Auchterarder in 1963 with a southbound van train. In 1959 the class totalled 842 engines, one of Britain's largest classes, with many examples now preserved. *Author*

Bank Hall (27A, closed 1966) had an allocation of 46 engines in 1950 and 41 in 1959. Eight L&YR 0-4-0STs were on the books in 1959 and one 2-4-2T, No 50721. The final allocation was for 17 engines which included four 'Jubilee' class 4-6-0s. The code changed from 23A to 27A in 1950 and from 27A to 8K in 1963.

Aintree (27B, closed 1967) had 55 engines in 1950 including 27 of the Fowler 0-8-0s. By 1959 the Fowlers had gone and had been replaced by 21 'WD' class 2-8-0s. The shedcode changed from 23B to 27B in 1950 and from 27B to 8L in 1963.

Southport (27C, closed 1966) was a six-road straight shed originating from L&YR days. The last L&YR 2-4-2T worked from here and the shed is now part of 'Steamport' museum. An L&YR 2-4-2T, No 1008, latterly BR No 50621, can be seen in the National Railway Museum. The Southport shedcode changed from 23C to 27C in 1950 and from 27C to 8M in 1963.

Wigan (27D, closed 1964) had an allocation of LMS and Standard types in 1959 with a total of 37 engines. The code changed from 23D to 27D in 1950 and from 27D to 8P in 1963.

Walton (27E, closed 1963) of CLC origin had 22 engines in 1959, including a few GCR types. The code changed from 13F to 27E in 1950 and from 27E to 8R in 1963.

Brunswick (27F, closed 1961) was a CLC shed originally with 33 engines in 1959 by which time LMS types had replaced the old GCR relics. The code changed from 13E to 8E in 1950 and from 8E to 27F in 1958. The five-road shed was extremely cramped and awkward to use.

Above:
The Fowler 0-8-0 class introduced in 1929 was classified '7F' by the LMS and was a development of the LNWR 'G2' class 0-8-0. No 49508 is seen ex-works at Horwich in June 1959 and came from a class of 175 engines built from 1929 to 1932. Aintree (27B) had an allocation of 27 engines in the class in 1959. *P. J. Hughes/Colour-Rail*

Below:
BR Standard Class 4MT 4-6-0 No 75027 was a Bank Hall (27A) engine in 1965 and subsequently became preserved in BR lined green livery as the engine was an ex-Western Region example. The class of 80 engines was introduced by BR in 1951 and No 75027 has been found a home on the Bluebell Railway. *Author*

Eastern Region

STRATFORD

30A	Stratford
	Chelmsford
	Enfield Town
	Southend (Victoria)
	Wood Street (Walthamstow)
30B	Hertford East
	Buntingford
30C	Bishop's Stortford
30E	Colchester
	Braintree
	Clacton
	Maldon
	Walton-on-the-Naze
30F	Parkeston

Sample Allocation 1959

Colchester 30E

Class 4 2-6-0	43152/3
Class 2 2-6-0	46468/69
Class B1 4-6-0	61000/61300/ 11/36/61/63/ 70/73
Class B17 4-6-0	61658/62/3/6/8
Class J19 0-6-0	64650/7/60/4/ 6/7
Class J15 0-6-0	65424/43/5/6/8/ 65/8/70/2/3
Class J17 0-6-0	65503/5/6/11/4/ 31/9/45/64
Class J69 0-6-0T	68552/73/9
Class N7 0-6-2T	69612/3/7/52/ 73/8/86/ 69727/32/3

Total 55 Engines

Stratford engine shed (30A, closed 1962), of the former Great Eastern Railway, had a truly massive allocation of engines, the 1950 list showing no less than 383 locos. There were four sub-sheds in the 1959 allocation with 197 engines. In 1950 apart from main line locomotives of the 'B1', 'B12' and 'B17' classes there were 105 'N7' class 0-6-2Ts and 64 'J67-J69' 0-6-0Ts all used on the intensive suburban steam-worked services until electrification. Stratford had the largest allocation in Britain and possibly the world.

Hertford East (30B, closed 1960) was a two-road shed of GER origin with 10 'N7s' allocated for suburban work in 1959.

Bishop's Stortford (30C, closed 1960) had an allocation of 11 engines in 1959, three 'J17' class 0-6-0s and eight 'L1' class 2-6-4Ts. The shed was in fact an open-air stabling point with coal, sidings and a turntable.

Colchester (30E, closed 1959) was a more conventional three-road straight shed with 55 engines in 1959 which included eight 'B1' class 4-6-0s and five 'B17s'.

Parkeston (30F, closed 1961) had 33 engines in 1959 and was a four-road shed with 16 'B1' class 4-6-0s, including five named examples.

The Great Eastern 'N7' class 0-6-2Ts worked the frequent steam suburban service out of Liverpool Street until electrification in 1961. No 69688 of Hertford East (30B) is seen at Buntingford in January 1959 on that long-forgotten branch line from St Margarets. *Author*

Above:
Stratford shed (30A) was the largest engine shed in the country and had an allocation of 21 'B12s' in 1950 out of a total of 383 engines. No 61545 was classified 'B12/3' by the LNER and was a Gresley rebuild of the original GER 4-6-0. *E. V. Fry/Colour-Rail*

Below:
The sole surviving 'B12' class 4-6-0 has taken many years to restore and in 1995 was returned to traffic on the North Norfolk Railway. No 61572 was an old Stratford engine and is seen here at Weybourne as LNER No 8572.
Alan C. Butcher

31 A CAMBRIDGE

31A	Cambridge
	Ely
	Huntingdon East
	Saffron Walden
31B	March
	Wisbech East
31C	King's Lynn
	Hunstanton
31D	South Lynn
31E	Bury St Edmunds
	Sudbury (Suffolk)
31F	Spital Bridge (Peterboro')

Sample Allocation 1959

Cambridge 31A

Class 4 2-6-0	43087
Class 2 2-6-0	46465–7
Class B1 4-6-0	61066/61104/71/82/61203/36/80/ 3/6/7/61301/14/60/71
Class B12 4-6-0	61577
Class B2 & B17 4-6-0	61607/8/13/14/16/23/39/44/ 51/2/61
Class K3 2-6-0	61817/34/49/80
Class E4 2-4-0	62785
Class J19 0-6-0	64646/54/8/61/73
Class J20 0-6-0	64683/95/6
Class J39 0-6-0	64803/64985
Class J15 0-6-0	65450/1/7/61/75/7
Class J17 0-6-0	65502/20/28/32/41/56/80/9
Class L1 2-6-4T	67701/12/13/8/20/1/2/3/33/4
Class J69 0-6-0T	68566/68609
Total 71 Engines	

Cambridge (31A, closed 1962) had a total of 71 engines allocated in 1959, a decline from the 101 of 1950. The seven-road straight shed was right next door to the station platform where observation was convenient. The shed was rebuilt by the LNER in 1932. The shed had 11 'B2'/'B17' 4-6-0s in 1959 as well as 14 'B1s'; the star attraction in 1959 was the GER 2-4-0 No 62785 the last 2-4-0 engine in use on BR. The locomotive was built at Stratford in 1891 for the GER and was of a class which once totalled 100 engines. The engine was restored to GER livery after withdrawal in 1959 and is now in the National Collection.

March (31B, closed 1963) had an allocation of 161 engines in 1950 and 131 in 1959. The two Whitemoor sheds served the marshalling yard; the 1950 allocation included 50 'WD' class 2-8-0s and 49 'K1'/'K3' class 2-6-0s. In the 1959 allocation the shed had five 'V2s' to supplement the 'B17' class 4-6-0s, the 'WDs' had declined to 18 examples and 'Britannia' class 4-6-2s had started to arrive.

King's Lynn (31C, closed 1959) had 47 engines in 1950 and 17 at closure of seven different classes. The shed, which was built by the GER, was of four roads and in use for servicing visiting engines until 1961.

South Lynn (31D, closed 1959) was the former Midland & Great Northern Joint Railway shed and had 17 engines in the final year of operation. Ivatt Class 4MT 2-6-0s were the dominant form of motive power in the final years of the M&GN and a member of the class, No 43106, has been preserved on the Severn Valley Railway.

Bury St Edmunds (31E, closed 1959) was a three-road brick and timber building of GER origin. Old GER types could be found here including 'E4' class 2-4-0s and 'F6' 2-4-2Ts.

Peterborough Spital Bridge (31F, closed 1960) was of Midland Railway origin and became part of the Eastern Region in 1950 when the code was altered from 16B to 35C. The shedcode was altered again in 1958 to 31F. The shed had an enclosed roundhouse Midland style and had 43 engines in 1950 and 34 in 1959. LNER types took over after 1950 but curiosities were the three ex-LT&SR type 4-4-2Ts reallocated there after replacement by more modern power on the LT&SR lines.

Top right:
Cambridge (31A) in the winter of 1959 with ex-Great Eastern 'J15' class No 65475 on snowplough duty. One member of the class, No 65462, has been preserved and can be seen on the North Norfolk Railway. *Author*

Centre right:
Ely with 'B2' class 4-6-0 No 61615 leaving in January 1959. The engine was a two-cylinder rebuild of the Gresley 'B17' class built in 1930 and rebuilt by Thompson in 1946. The engine was allocated to Cambridge (31A) from 1956 and was withdrawn in February 1959. *Author*

Botton right:
Class D16/3 4-4-0 No 62613 is seen at March in April 1960. The 'D16/3' class was a rebuild of the 'D16/2' the former 'Claud Hamilton' class of the GER. No 62613 was a Peterborough engine (31F) in 1959. *A. N. H. Glover*

32A	Norwich Thorpe
	Cromer Beach
	Dereham
	Wymondham
32B	Ipswich
	Felixstowe Town
	Stowmarket
32C	Lowestoft Central
32D	Yarmouth South Town
32E	Yarmouth Vauxhall
32F	Yarmouth Beach
32G	Melton Constable
	Norwich City

Sample Allocation 1959

Norwich 32A

Class 4 2-6-0	43145/6/56/60/1
Class B1 4-6-0	61042/3/5/6/8/61223/35/70/79/61312/17/99
Class B12 4-6-0	61514/30/3/68/71
Class B17 4-6-0	61636/54
Class K3 2-6-0	61826/77/61908/18/39/49/53/7/70/1/3/81/9
Class D16 4-4-0	62511/7/24/40/4
Class J19 0-6-0	64641/3/4/74
Class J39 0-6-0	64731/61/64900/13
Class J15 0-6-0	65469/71
Class J17 0-6-0	65519/42/51/3/7/66/70/81/6
Class L1 2-6-4T	67714/7/86
Class J68 0-6-0T	68640/1/5
Class J50 0-6-0T	68899/68905/24
Class N7 0-6-2T	69107
Class 7P/6F 4-6-2	70000–3/05–13/30/4/6–41
Class WD 2-8-0	90559

Total 92 Engines

The four-road brick-built shed at Norwich (32A, closed 1962) certainly had a variety of motive power from old GER 'B12s' through LNER types to modern 'Britannias'. It was the proud boast of the late Bill Harvey, the Norwich shedmaster, that he had 21 'Britannia' class Pacifics allocated to his shed to work the expresses to Liverpool Street. Two of the 'Britannias' have survived into preservation — Nos 70000 *Britannia* and 70013 *Oliver Cromwell*.

Ipswich (32B, closed 1959 to steam) had an allocation of 90 engines in 1950 and 50 in 1959. The old GER shed was completely rebuilt in 1954 by BR in anticipation of dieselisation. The 1959 allocation included 12 'B1', three 'B12' and eight 'B17' class 4-6-0s as well as GER and LNER 0-6-0s used for freight work and five 'L1' class 2-6-4Ts. The surviving 'B12', No 61572, only recently restored to operational condition and LNER green, was an Ipswich engine, having been built in 1928 by Beyer Peacock to the GER design. The engine was withdrawn from BR service in 1961.

Lowestoft (32C, closed 1960) was a four-road GER shed with 37 engines in 1950 and 18 in 1959. The allocation included two 'B17' class 4-6-0s as well as ex-GER types.

Yarmouth South Town (32D, closed 1959) had 22 engines in 1950 and six in 1959. The allocation was three 'B17s', two 'D16s' and one 'J68' 0-6-0T in 1959. The shed was rebuilt by BR for dieselisation, but the line served closed in May 1970.

Yarmouth Vauxhall (32E, closed 1959) was a magnificent Great Eastern building of two roads with end doors and a clerestory roof — ideal for modellers. In 1995 the shed housed seven 'D16' 4-4-0s, one 'F4' 2-4-2T and one 'J67' 0-6-0T — all ex-GER types.

Yarmouth Beach (32F, closed 1959) was the M&GN building and housed Ivatt Class 4 2-6-0s at closure.

Melton Constable (32G), with an allocation of 26 engines in 1950, was at the centre of the M&GN system, most of which, including the shed, closed from 1 March 1959. The Ivatt Class 4 2-6-0s were relocated after closure and very little remains to be seen of the station site today.

A vintage scene at Hopton-on-Sea with a 'D16/3' class 4-4-0 No 62511 with the 10.15am Liverpool Street to Gorleston in 1957. The 4-4-0 was a Norwich (32A) engine at the time and is seen with a Saturday-only summer special. *E. Alger/Colour-Rail*

Above:
Norwich (32A) had the Great Eastern lines' entire allocation of 21 'Britannia' Pacifics in 1959 and worked the main line expresses until electrification. The 'Britannia' class Pacifics were re-allocated and No 70013 ended its final days at Lostock Hall to work the last BR steam-hauled train in 1968. *J. Phillips*

Below:
The restored Great Eastern 'J17' class 0-6-0 No 65567 is seen in store at Hellifield in 1965. The engine has been restored to LNER livery and is the sole surviving example of the Holden GER design. The engine was allocated to Lowestoft Central (32C) in 1959. *A. N. H. Glover*

33 A ⃝ PLAISTOW

33A	Plaistow
33B	Tilbury
33C	Shoeburyness

Sample Allocation 1959
Plaistow 33A
Class 3F 0-6-2T	41981
Class 4 2-6-4T	42226/7/54/5/7
Class 3F 0-6-0T	47262/47312/28/51/47484/47512/55
Class J39 0-6-0	64951/2/3/4/6/7/8/62/5/8
Class 4 2-6-4T	80096–105/80131–36
Class WD 2-8-0	90196/90256/90653
Total 42 Engines	

Plaistow shed (33A) in 1958 with a line of old LT&SR tanks of the 0-6-2 and 4-4-2 type. The 0-6-2Ts were a Whitelegg design dating from 1903 and known as the 'G9' class, some being built under MR auspices. One Atlantic tank, No 41966, has been restored to original condition as No 80 *Thundersley. E. V. Fry/Colour-Rail*

Plaistow (33A, closed 1962), with an allocation of 83 engines in 1950 and 42 in 1959, was the principal LT&SR shed. The shed was recoded from 13A in 1949. The establishment became a sub-shed of Tilbury in 1959 and consisted of an eight-road straight building in brick. The old LT&SR 4-4-2Ts were replaced by BR Standard types in the 1950s and out of the batch of 16 Standard Class 4 2-6-4Ts shown no less than seven have been preserved — Nos 80097/8/80100/4/5/35/6 — quite a record! The Class 4 standard 2-6-4T is an ideal engine for the preserved railways as it is of medium power and does not need to be turned when in use, although the survival of so many of the class stems from the engines' survival for so long at the Woodham Bros scrapyard at Barry.

Tilbury (33B, closed in 1962) was a modern four-road straight shed situated within the triangle that led to Tilbury Riverside station (recently closed). The 1959 allocation consisted of 12 Class 4 2-6-4Ts and seven 'WD' class 2-8-0s. Out of the 12 Class 4 2-6-4Ts four — Nos 80072/8/9/80 — have been preserved. The shedcode changed from 13C to 33B in 1949 after the Eastern Region takeover.

Shoeburyness (33C, closed 1962) had an allocation of 48 Class 4 2-6-4Ts in 1959 and consisted of a four-road straight shed adjacent to the station. No Class 4 standards were allocated there in 1959 but the shed had the entire batch of Stanier three-cylinder 2-6-4Ts allocated to work the fast passenger services to Fenchurch Street. The shedcode was altered from 13D to 33C in 1949 when the Eastern Region took over from the London Midland.

Above:
The London, Tilbury & Southend line was well known for the 4-4-2Ts of the Whitelegg '79' class. The Midland and LMS railways perpetuated the Atlantic tank design, and the class survived into BR days. No 41978 is seen at Shoeburyness (33C) in 1958 shortly before withdrawal from service. *T. B. Owen/Colour-Rail*

Below:
Class 4 2-6-4 tank No 80069 leaves Pantyffynnon on the Central Wales line with a stopping passenger train. The engine (ex-33B) was one of a batch of Class 4 tanks sent to the LT&SR lines to replace the old Atlantic tanks but was then displaced by electrification. *Author*

34A	King's Cross
34B	Hornsey
34C	Hatfield
34D	Hitchin
34E	New England
34F	Grantham

Sample Allocation 1959

King's Cross 34A

Class A4 4-6-2	60003/6/7/8/10/3/4/5/7/21/2/5/6/8/9/30/2/3/4
Class A3 4-6-2	60039/44/55/9/61/2/6/60103/9/10
Class V2 2-6-2	60800/14/20/54/62/71/60902/3/14/50/83
Class B1 4-6-0	61075/61174/9/61200/72/61331/64/93/4
Class L1 2-6-4T	67757/67/8/70/2/3/4/6/9/80/3/4/7/92/3/7/67800
Class J52 0-6-0ST	68846
Class J50 0-6-0T	68946
Class N2 0-6-2T	69490/2/8/69504/6/12/15/17/20/1/3/4/6/8/9/32/5/8/9/41/3/5/6/9/68/70/4/5/6/8/9/80/1/3/4/5/9/92/3

Total 107 Engines

King's Cross (34A, closed 1963) had a fascinating selection of engines which worked the East Coast main line and the all-steam suburban services of the former Great Northern Railway. The 1959 allocation was for 107 engines. Some famous engines were resident here including No 60007 *Sir Nigel Gresley*, No 60022 *Mallard*, No 60103 *Flying Scotsman* and No 60800 *Green Arrow*, all of which are preserved.

Hornsey shed (34B, closed 1961) had a nearly all-tank engine allocation of 81 in 1950 and 58 in 1959. 'J52' class 0-6-0STs gave way to 'J50' class 0-6-0Ts by 1959, the engines being used for shunting and empty stock working.

Hatfield (34C, closed 1961) had an allocation of 28 engines in 1950 and 21 in 1959. The all-tank allocation there consisted of 'N2' and 'N7' class 0-6-2Ts. The brick-built shed only had one through road and most of the engines which worked the steam passenger services were stabled out in the open.

Hitchin shed (34D, closed 1961) was a two-road ex-GNR building with a wooden roof situated behind the up passenger platform. The 1959 allocation of 29 engines, which had declined from the 33 of 1950 included eight 'B1' class 4-6-0s, nine Thompson 'L1' class 2-6-4Ts, seven GNR 'J6' 0-6-0s, three 'N2' 0-6-2Ts, one 'J68' 0-6-0T and a solitary GER 'J15' 0-6-0, which was much used on the line to Bedford.

New England (34E, closed 1965) was rebuilt by BR in 1952 and had a large allocation of 111 engines, which was a considerable reduction from the 213 of 1950. The loco allocation included 'A2' class Pacifics, 'V2' class 2-6-2s and a large selection of 'WD' class 2-8-0s, 53 in 1950 and 20 in 1959 which were supplemented by 25 '9F' 2-10-0s. A feature of the shed was the overhead watering gantry of United States/South African Railways style. There was no turntable, the locomotives having to turn on the shed triangle.

Grantham (34F, closed 1963) had 35 locomotives in 1950 and 41 in 1959 which included Gresley 'A3' Pacifics (12) and Thompson 'A2' class 4-6-2s (three). Two GCR 'A5' 4-6-2Ts were also on the list. A unique feature introduced in 1951 was the turning triangle which replaced two turntables and featured a scissors crossover. There were two sheds: 'old' and 'new', both of four roads.

Flying Scotsman was a King's Cross (34A) engine in 1959 and this famous engine has travelled far during its distinctive career which has included visits to the USA and Australia. The engine has roamed all over Britain and is seen here at Eastleigh in September 1963. *J. Phillips*

Above:
Klng's Cross (34A) had 17 'A4' class Pacifics in 1959 and No 60033 *Seagull* is seen passing Hitchin in December 1961. No less than four examples from the shed's 1959 list are preserved including the famous No 60022 *Mallard*.
J. Phillips

Below:
The King's Cross suburban scene is reincarnated with 'N2' class No 69523 posing as No 69568 on a van train at Rothley on the GCR. 'N2' classes worked suburban steam trains before the GNR electrification. Nos 69568 and No 69523 were King's Cross (34A) engines in 1959. *Author*

49

DONCASTER

36A	Doncaster
36C	Frodingham
36E	Retford

Sample Allocation 1959
Retford 36E

Class B1 4-6-0	61126/61208/11/12/31
Class O4 2-8-0	63608/37/55/88/63736/82/5/63818/
	63914
Class O2 2-8-0	63924–27/37/44/5/7/9/59/61/5/70/1/
	2/6/9/80/2/6/87
Class J6 0-6-0	64174/8/88/64234/6/45
Class J11 0-6-0	64280/3/7/64321/85/95/64421/3/50/1
Class J39 0-6-0	64714/59/64830/82/93/8/64906/8/70
Class N5 0-6-2T	69322
Total 61 Engines	

Doncaster (36A) was the former Great Northern Railway's largest shed with a healthy allocation of Pacifics as well as old GNR types in BR days. The allocation included 23 'V2' class 2-6-2s and No 60870 is seen at Sleaford in 1961 working the 12.54pm Doncaster to March. *Author*

Doncaster (36A, closed 1966) was the principal GNR shed and was a 13-road straight shed with a turning triangle. The 1950 allocation included 180 engines, and the 1959, 191. The 1959 allocation included seven 'A3s', 13 'A1s', 23 'V2s' and 27 'B1s'. Old GNR 2-8-0s in the 'O2' class amounted to 31 engines and the 12 'WD' class 2-8-0s were supplemented by 16 class '9F' 2-10-0s. The solitary 'W1' class 4-6-4 (withdrawn in June 1959) the 1937 rebuild of Gresley's four-cylinder compound with water tube boiler of 1929 was also a Doncaster engine. Doncaster shed was the last steam shed to be in use on the Eastern Region.

Frodingham (36C, closed 1966) was an ex-GCR shed of five roads. The locomotives were all freight types and included 28 ex-GCR 'O4s', seven 'J11s', two 'J50s' and the five 'Q1' class 0-8-0Ts of Robinson GCR origin. There were 29 'WD' class 2-8-0s and seven '9F' 2-10-0s. The 'Q1' class, built at Gorton from 1902 as GCR Class 8A, was extinct with the withdrawal of No 69936 in September 1959. The engine had been rebuilt to 'Q1/2' class by Thompson in 1945.

Retford (36E, closed 1965) had two sheds, one ex-GNR and one ex-GCR. The GCR shed closed in January 1965 to make way for the flyunder which removed the flat crossing of the East Coast main line by the GCR line.

Top:
Gresley 'O2' class No 63926 of 36E (Retford) is seen working an empty mineral train near Ordsall in June 1959. The 'O2' class 2-8-0s were three-cylinder engines introduced in 1921 for main line freight work and were classified '8F' by BR. None of the class were preserved.
P. J. Hughes/Colour-Rail

Above:
Robinson Great Central 2-8-0 No 63601 has been preserved and was one of a class of numerous freight engines produced from 1911 onwards, many being rebuilt by the LNER. The engines were known as the 'O4' class on the LNER, a few of them being rebuilt to Class O1 under Thompson's regime. No 63601, an ex-Frodingham (36C) engine, is seen here in undercoat paint *en route* for preservation. *Author*

⬭ 40 A LINCOLN

40A	Lincoln
40B	Immingham
	Grimsby
	New Holland
40E	Colwick
40F	Boston
	Sleaford
	Spalding

Sample Allocation 1959

Lincoln 40A

Class 1P 0-4-4T	58065
Class B1 4-6-0	61009/26/61202/48/58/61405
Class K3 2-6-0	61802/6/7/28/48/59/89/94/61919/44/60
Class J6 0-6-0	64207/19/78
Class J11 0-6-0	64318/71/64430
Class J39 0-6-0	64726/34/41/51/5/64881/9/90/6/ 64937/59/60/1/6/84
Class L1 2-6-4T	67769
Class J69 0-6-0T	68501/10/28/43/60/81/99
Class A5 4-6-2T	69803/8/20/21
Total 51 Engines	

Lincoln (40A, closed 1964) was of GNR origin and was a four-road straight shed rebuilt by the Eastern Region. The 1959 allocation of 51 engines includes a solitary Midland '1P' 0-4-4T of 1889. GNR 0-6-0s in the 'J6' class and GCR 0-6-0s in the 'J11' class were represented as well as four surviving GCR 'A5' class 4-6-2Ts.

Immingham (40B, closed 1966) was of GCR origin and had 120 engines in 1950 and 94 in 1959. The 1950 allocation included eight 'Director' class 4-4-0s in the 'D11' class. No 62660 *Butler Henderson* built in 1920 of a class of 11 engines was an Immingham engine and is now in the National Collection, having worked for some years on today's preserved Great Central Railway. By 1965 the allocation at Immingham had fallen to 27 engines, of which 13 were 'B1s', nine were 'WDs' and five were '9F' 2-10-0s.

Colwick (40E, closed 1966) was a huge shed of GNR origin and had an allocation of 199 engines in 1950 and 147 in 1959. The shed was a dead-end straight brick building with 18 roads. The allocation of 1959 reflected the shed's status as a provider of freight engines for the Nottinghamshire coal traffic. There were 21 members of the 'O1/O4' class and 45 of the 'WD' class 2-8-0. The Colwick shedcode changed from 38A to 40E in 1958 and to 16B in 1966. Five ex-GCR 4-6-2Ts — Nos 69800/5/9/12/25 — were also on the 1959 list.

The ex-GNR shed at Boston (40F, closed 1964) consisted of three buildings and in 1959 the majority class was the Ivatt Class 4 2-6-0, of which there were 27 out of the total allocation of 51 engines.

A picturesque scene depicting ex-Great Northern 0-6-0 No 64172, a 'J6' class, at Spalding Town on the last day of services on the former Midland & Great Northern Joint Railway. Boston (40F) shed had 11 'J6' 0-6-0s on the 1959 list used for local freight and passenger work. None of the class was preserved. *T. J. Edgington/Colour-Rail*

Below:
Southwell was a Midland Railway station and MR architecture as well as signalling is in evidence in this May 1959 photograph. Ex-GCR 'A5/1' class 4-6-2T No 69808 is seen here with a single coach. The engine was shedded at Lincoln (40A) in 1959 and the line closed on 15 June of that year. *B. Hilton/Colour-Rail*

Bottom:
Sheffield Victoria of late Great Central fame sees 'K2' class 2-6-0 No 61778 with empty stock in September 1958. Gresley's 'K2' class of 1914 was used on cross-country and branch lines and Immingham (40B) shed had 10 examples in 1959. *G. Warner/Colour-Rail*

SHEFFIELD DARNALL

41A	Sheffield (Darnall)
41B	Sheffield (Grimesthorpe)
41C	Millhouses
41D	Canklow
41E	Staveley (Barrow Hill)
41F	Mexborough Wath
41G	Barnsley
41H	Staveley (ex-GCR)
41J	Langwith
41K	Tuxford

Sample Allocation 1959
Sheffield Darnall 41A

Class B1 4-6-0	61033/41/4/7/50/1/83/61105/38/50–54/ 62/9/81/3/61313/5/6/27/34
Class K2 2-6-0	61728/47/60/1
Class K3 2-6-0	61816/25/61907/38/43/67
Class D11 4-4-0	62660/1/2/4–70
Class O4 2-8-0	63574/83/99/63604/9/21/4/40/5/58/61/ 80/5/95/63710/33/4/7/42/8/71/83/ 63821/2/46/50/2/81/2/8/9
Class J11 0-6-0	64329/73/87/94/64419/41/3/5/7
Class J39 0-6-0	64702/19/36/46/64804/7/8/78
Class N5 0-6-2T	69258/90/4/6/69314/61
Total 97 Engines	

Sheffield Darnall (41A, closed 1963) was built by the GCR and had 97 engines in 1959, mainly of GCR origin. The 10 'Directors' included the now preserved 'D11' *Butler Henderson*, which was withdrawn from normal service in October 1960. The last 'D11' 4-4-0 No 62666 *Zeebrugge* was withdrawn in December 1960, after 38 years' service. The Darnall code changed from 39B to 41A in 1955.

Sheffield Grimesthorpe (41B, closed 1961) was of Midland origin and consisted of an internal roundhouse in the MR style. The 1959 allocation consisted of 56 engines of Midland or LMS origin. The 1950 list shows 80 engines. The code was changed from 19A to 41B in 1958, when the Eastern Region took over.

Millhouses (41C, closed 1962) was another shed of Midland origin with 41 engines in 1950 and 33 in 1959. The 1959 allocation included 12 'Jubilee' class 4-6-0s, nine Class 5 Standard 4-6-0s and four Class 2 Standard 2-6-0s. The code changed from 19B to 41C in 1958.

Canklow (41D, closed 1965) was another Midland enclosed roundhouse with 53 engines in 1950 and 45 in 1959. MR, LMS and Standard types predominated, and the code was changed from 19C to 41D in 1958.

Staveley (41E, closed 1965) was also known as Barrow Hill and consisted of an MR roundhouse. MR types included the '0F' class 0-4-0T of 1907 of which the shed had four examples. The code changed from 18D to 41E in 1958.

Mexborough (41F, closed 1964) was of GCR origin and consisted of a 15-road straight layout. The 1950 allocation was for 119 engines and the 1959 allocation was for 87. In 1950 there were 56 'WD' class 2-8-0s which had been reduced in number to 40 by 1959. There were also 30 'O4' 2-8-0s in 1959, the code having changed from 36B to 41F in 1958.

Barnsley (41G, closed 1960) of GCR origin was a two-road shed at Exchange station. There were 40 locomotives in 1950 and 37 in 1959, mainly ex-GCR types including the graceful Robinson 'C14' class 4-4-2Ts.

Staveley (41H, closed 1965) was a five-road shed of the former GCR with 34 engines in 1950 and 37 in 1959. The shed was rebuilt by BR in 1952 and the 1959 allocation included a solitary 'Director' class No 62663. The code changed with Regional reorganisation in 1958 from 38D to 41H.

Langwith (41J, closed 1966) was of GCR origin and consisted of two sheds side by side. The 1950 allocation was for 61 engines and the 1959 list showed 71, of which 35 were GCR 'O4' class 2-8-0s. The code changed from 40E to 41J in 1958.

Tuxford (41K, closed 1959) on the LD&ECR had an allocation of 15 engines in 1950, all ex-GCR. The code changed from 40D to 41K in 1958.

Top right:
Old Great Central 0-6-0 No 64346 is seen on Lincoln shed in April 1960. The GCR 'J11' 0-6-0s were known to railwaymen as 'Pom Poms' after the noise they made when working. No 64346 was a 'J11/3' class having been rebuilt by Thompson under the LNER. The engine was based at Langwith (41J) in 1959. *T. A. Murphy/Colour-Rail*

Centre right:
Nottingham Victoria with station pilot No 62668 *Jutland*, a Great Central 'Director' class allocated to Sheffield Darnall (41A) in 1959. Sheffield Darnall shed had 10 'Directors' in 1959 and was closed in 1963. *G. Warner/Colour-Rail*

Bottom right:
Old Midland 0-6-0 tank No 41835 with half cab, round-top boiler and Salter safety valves is seen in 1958 at Rotherham. Classified as '1F' by the LMS, the engines originated from the Johnson 1878 design. The engine was shedded at Canklow (41D) at the time. *P. J. Hughes/Colour-Rail*

North Eastern Region

(50 A) YORK

50A	York
50B	Leeds (Neville Hill)
50C	Selby
50D	Starbeck
50E	Scarborough
50F	Malton Pickering
50G	Whitby

Sample Allocation 1959

York 50A

Class 2 2-6-2T	41252
Class 4 2-6-4T	42083/5
Class 2 2-6-0	46480/1
Class 3F 0-6-0T	47239/54/47334/47403/18/21/36/48/56/ 47607
Class A1 4-6-2	60121/38/40/6/53
Class A2 4-6-2	60501/2/3/12/15/22/4/6
Class V2 2-6-2	60828/37/9/47/55/6/64/76/7/8/87/95/ 60907/18/25/39/41/6/54/60/1/3/8/ 74/5/7/81/2
Class B1 4-6-0	61002/53/69/71/84/6/61288/61337
Class B16 4-6-0	61410/3/6–24/6/30/34–41/3/4/8–55/7/ 60–65/7/8/72/3/5–7
Class K1 2-6-0	62046/8/9/50/6/7/61/2/3
Class D49 4-4-0	62740
Class J25 0-6-0	65698/65714
Class J27 0-6-0	65845/74/83/7/90/4
Class J77 0-6-0T	68392/68431
Class J94 0-6-0ST	68032/40/46/61
Class J71 0-6-0T	68309
Class J72 0-6-0T	68677/87/68736/69016/20
Class 3 2-6-0	77012
Class WD 2-8-0	90068/90200/30/6/90405/24/45/75/ 90543/78

Total 150 Engines

Sheds at York dated from the earliest times of the railway and included workshops. The first roundhouse was built in 1850 followed by a second in 1852 and a third in 1864. York shed (50A, closed 1967) under BR consisted of York North shed with a quadruple roundhouse and York South with two separate roundhouses which were demolished in 1963. The quadruple roundhouse (now part of the NRM) was rebuilt by BR in 1958 when two of the roundhouses were rebuilt into a straight shed of five roads for diesel use. The 1950 allocation showed five 'A1' and six 'A2' class Pacifics, with 30 'V2s' and 10 'D49' 4-4-0s out of a total of 174 engines. In 1959 the 47 'B16' class 4-6-0s of 1950 had declined slightly in number to 45 engines but the 'D49s' had gone down to only one example. The 1959 list shows 150 engines of 18 different classes.

Neville Hill shed at Leeds (50B, closed 1966) was of NER origin and contained four enclosed roundhouses. BR rebuilt the shed in 1958 for diesel use and demolished two of the roundhouses. The code changed from 50B to 55H in 1960. The 1950 allocation shows 81 engines. Four 'A3' class 4-6-2s, 10 'B1' 4-6-0s and 15 'B16' class 4-6-0s were amongst the 54 engines on the 1959 list.

Selby (50C, closed 1959) had 42 engines in 1959 for its two internal roundhouses. In 1950 the shed had an all-NER allocation with the addition of four Sentinel 0-4-0s. The 1959 allocation shows a decline in NER types, but six 'B16' class 4-6-0s were still on the list.

Starbeck (50D, closed 1959) was a two-road shed which had been rebuilt by BR, and had an allocation of 30 engines in 1959. The 1950 list of 45 engines shows 12 'D49' class 4-4-0s, which had declined in number to six by 1959.

Scarborough (50E, closed 1963) had 14 engines in 1959 which included three 'Shire' 'D49' class 4-4-0s. The shed consisted of two straight buildings and could accommodate many visiting engines during the summer months. The shed was rebuilt by BR to accommodate diesels.

Malton (50F, closed 1963) was a two-road building in brick and had 13 engines in 1959 including two 'A8' class 4-6-2Ts.

Whitby (50G, closed 1959) had 13 locos in 1950 and five by 1959. The 1950 list shows seven 'A8' 4-6-2Ts at the two-road shed.

Top:
'A2' class No 60524 *Herringbone* was a York engine in 1959 and was classified 'A2/3' by the LNER. York (50A) had eight 'A2' class 4-6-2s for East Coast express work but these were transferred to other depots when main line diesels took over. No 60524 is seen at Dunning with an Aberdeen to Glasgow train in 1963. *Author*

Above:
Gresley 'J39' class 0-6-0 No 64942 is seen at Waterhouses with a Durham Gala excursion in 1960. The engine was shedded at Starbeck (50D) in 1959 and the branch closed to all traffic on 28 December 1964. The station site is now a park. *Author*

Right:
The North Eastern two-cylinder 0-8-0s became LNER Class Q6 and one example has been restored on the North Yorkshire Moors Railway. No 63395 was shedded at Selby (50C) in 1959 and is seen at Eller Beck restored to its original NER livery. *Author*

51 A DARLINGTON

51A	Darlington	
	Middleton in Teesdale	
51C	West Hartlepool	
51E	Stockton	
51F	West Auckland	
51G	Haverton Hill	
51J	Northallerton	
51L	Thornaby	

Sample Allocation 1950
Northallerton 51J

Class D20 4-4-0	62347/59/88/91
Class J21 0-6-0	65030
Class J25 0-6-0	65645/93/65725
Class G5 0-4-4T	67324/44/6
Class Y3 0-4-0T	68159
Class N1 0-6-2T	69101
Total 13 Engines	

Sample Allocation 1959
Northallerton 51J

Class 2 2-6-0	46471
Class K1 2-6-0	62044
Class 2 2-6-0	78010/11/2/4/5
Total 7 Engines	

Darlington (51A, closed 1966) had two buildings at Darlington (Bank Top): one roundhouse dating from the 1860s and a straight nine-road shed built by the LNER in 1939. The 1950 allocation consisted of 112 engines and the 1959 allocation of 70. In the final year of operation (1965) the number had come down to 37 locomotives. Two 'A3' class Pacifics, one 'V2' 2-6-2 and 24 'B1' 4-6-0s headed the list in 1959 with a solitary 'A8' 4-6-2T, No 69887.

West Hartlepool (51C, closed 1967) had a straight shed of three roads and a roundhouse, the 1959 allocation being for 57 engines, mainly of NER origin.

Stockton (51E, closed 1959) had 54 engines in 1950 and 31 in 1959, the year of closure. The shed, which was of NER origin, was an eight-road straight arrangement with an allocation of 13 different classes in 1950, including two 'A7' 4-6-2Ts, an 'A8' 4-6-2T and a 'T1' 4-8-0T, as well as 11 'B1' 4-6-0s and 15 'WD' 2-8-0s.

West Auckland (51F, closed 1964) was an enclosed roundhouse built by the NER, with 40 engines in 1950 and 35 in 1959 with a healthy representation of NER classes.

Haverton Hill (51G, closed 1959) was a four-road straight NER shed with 21 engines allocated in 1950 and 1959, mainly 'Q6' 0-8-0s but five 'B1' 4-6-0s and three 'WD' 2-8-0s appeared on the 1959 list.

Northallerton (51J, closed 1963) was a perfect two-road brick-built shed with a slate roof and end opening wooden doors. The sample allocation shows an all-NER list in 1950 (except the 'Y3' and the 'N1'). The allocation had changed completely by 1959 and all the ex-NER types had been replaced.

Thornaby (51L) was opened as a new shed in June 1958 and was built to be converted for diesel use. The shed consisted of a modern straight building and a roundhouse. The shed replaced Newport (51B) and Middlesbrough (51D) and is still in use. The 1959 steam allocation included 30 'Q6' 0-8-0s and 27 'WD' class 2-8-0s. Steam finished at Thornaby in 1964.

Above right:
A vintage unfitted freight headed by North Eastern 'P3' class 0-6-0 No 2392 rolls into Goathland on the NYMR. The engine ended its BR days as No 65894 at Tyne Dock (52H), having been a Thornaby (51L) loco. The locomotive is seen in North Eastern Railway livery although it was built after the Grouping. *Author*

Centre right
A not-too-clean 'K1' class 2-6-0 No 62065 is seen on the York breakdown train in June 1965 with the Minster in the background. The engine was shedded at Stockton (51E) in 1959, being one of the eight engines at that depot at the time. *K. R. Pirt/Colour-Rail*

Bottom right:
No 62005, the now preserved 'K1' class, was a Darlington (51A) engine in 1959 and was built in 1949 to the LNER Peppercorn 2-6-0 design. The engine was the only member of the class to have been preserved and can be seen today on the North Yorkshire Moors Railway. *Author*

52 A GATESHEAD

52A	Gateshead
	Bowes Bridge
52B	Heaton
52C	Blaydon
	Alston
	Hexham
52D	Tweedmouth
	Alnmouth
52E	Percy Main
52F	North and South Blyth
52G	Sunderland
	Durham
52H	Tyne Dock
	Pelton Level
52J	Borough Gardens
52K	Consett

Sample Allocation 1959
Gateshead 52A

Class A4 4-6-2	60001/2/5/16/18/19/20/23
Class A3 4-6-2	60038/40/2/5/51/2/60/70/1/5/6/ 8/91
Class A1 4-6-2	60115/24/9/32/5/7/42/3/5/7/50/1/4/5
Class A2 4-6-2	60516/8/21/38
Class V2 2-6-2	60805/7/9/33/60/8/60904/23/29/34/40/2/7/9/52/64/7/79
Class B1 4-6-0	61012/22
Class J39 0-6-0	64704/64852/65/9
Class J25 0-6-0	65656/65700
Class V1 & V3 2-6-2T	67637/9/57/87–90
Class J71 0-6-0T	68263/83/68314
Class J72 0-6-0T	68674/5/80/93/68720/3/31/44/69001/5/27
Class N10 0-6-2T	69097/69109

Total 88 Engines

Gateshead (52A, closed 1965) was an enclosed shed with four roundhouses of NER origin with a straight three-road shed for Pacific classes which were too big for the roundhouses' turntables. In 1956 BR rebuilt two of the turntables and provided accommodation for the largest engines. In 1964 the shed was rebuilt into a straight road building for diesel traction. The 1950 allocation was for 91 engines but by 1965 the number had fallen to 18. With 39 Pacifics allocated in 1959 the establishment was well worth a visit. One of the shed's 'A4s', No 60019 *Bittern*, has survived into preservation.

Heaton (52B, closed 1963) was another big NER passenger shed, situated north of the Tyne. The 1950 allocation showed 119 engines which declined to 85 by 1959. The allocation in 1959 included eight 'A3' class 4-6-2s, six 'A1' 4-6-2s and 19 'V2' class 2-6-2s.

Blaydon (52C, closed 1963) had 79 locos in 1950 and 52 in 1959. The shed, which opened in 1900, was of the NER enclosed roundhouse type with two turntables. The allocation was mainly ex-NER or ex-LNER freight types.

Tweedmouth (52D, closed 1966) had 47 engines in 1950 and 37 in 1959, the final allocation being just eight engines in 1965. Tweedmouth had a roundhouse and a four-road straight shed. Two 'A3' class Pacifics and five 'B1' 4-6-0s were on the shed's list in 1959. The old NER 'D20' class 4-4-0s were based here in the early 1950s to work local passenger trains.

Percy Main (52E, closed 1965) of NER origin had two straight sheds, one through and the other a dead end. The allocation was entirely of 'J27' class 0-6-0s, there being 24 in 1950 and 21 in 1959.

North and South Blyth (52F, closed 1967) were two separate sheds on either side of the river, a unique arrangement on BR. The mainly ex-NER engine stock fluctuated over the years from 26 in 1950 to 20 in 1959 and 27 in 1965. North Blyth was an NER roundhouse and South Blyth was a straight six-road arrangement.

Sunderland (52G, closed 1967) had 57 engines in 1950, 45 in 1959 and 23 in 1965. The shed was recoded in 1958 from 54A and consisted of a roundhouse and a straight shed later rebuilt by BR. The scene of the last knockings of steam on the Region, the shed closed in September 1967.

Tyne Dock (52H, closed 1967) had 48 engines in 1950 and 44 in 1959, the shed consisting of a roundhouse and a straight shed side by side. This freight-only shed closed on 9 September 1967 and the occasion saw the end of steam on the Region. Locomotives Nos 65894, 63395 and 62005 were later to be preserved, the latter being the last main line steam locomotive to work in the northeast. The shed was recoded from 54B to 52H in 1958.

Borough Gardens (52J, closed 1959) had an NER-type internal roundhouse with four turntables, an allocation of 47 engines in 1950 and 44 in 1959. Old NER types prevailed and the shed was recoded from 54C to 52J in 1958.

Consett (52K, closed 1965) was an NER straight two-road shed with 13 engines in 1950, 14 in 1959 and eight in 1965. The shedcode was altered from 54D to 52K in 1958. Nothing remains to be seen at Consett today as the whole railway site has been razed to the ground.

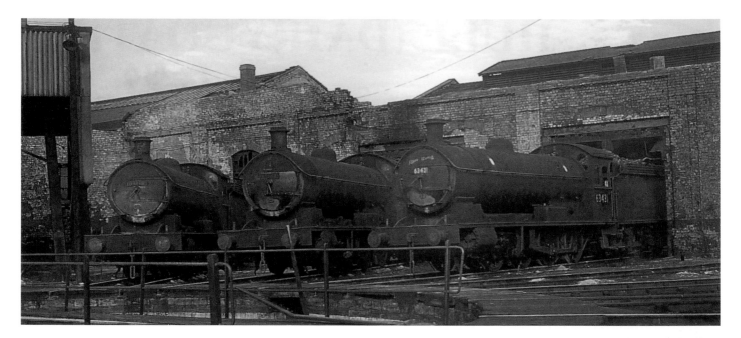

Above:
Tyne Dock shed (52H) in the last days of steam with 'Q6' class 0-8-0s in 1965. Engines 63389, 63384 and 63431 of the North Eastern Railway were the Raven two-cylinder 0-8-0s used for mineral trains and were amongst the last pre-Grouping engines to work on BR. *A. N. H. Glover*

Right:
Newcastle Central station pilot was a 'J72' class 0-6-0T and No 68723 can be seen here resplendent in its NER green livery in July 1960. The station pilots at Darlington and York were also restored to NER colours with the BR and NER crest on the tank sides. The NER design dated from 1898 with the engine being allocated to Gateshead (52A) shed. *Author*

Below:
Gateshead (52A) 'A1' No 60142 *Edward Fletcher* crosses the bridge at Berwick-upon-Tweed with an up train in February 1962. The engine was withdrawn in 1965 and no examples of the Peppercorn 'A1' class have survived into preservation, although a new one is being built. *J. Phillips*

53 A HULL (DAIRYCOATES)

Sample Allocation 1959
Hull Botanic Gardens 53B

Class B1 4-6-0	61010/68/80/61215/89/ 61304/05/06
Class D49 4-4-0	62701/7/10/17/20/2/3/60
Class V1 & V3 2-6-2T	67635/8/40/63/77/82/4/6
Class J73 0-6-0T	68363
Class 3 2-6-0	77001

Total 26 Engines

The three sheds at Hull reflected the importance of the area for freight traffic. The shed at Hull (Dairycoates) (53A, closed 1967) was the largest on the North Eastern Region. It dated from 1913 and was probably the largest covered shed on BR. With six internal turntables and a straight shed, all engines could be accommodated under cover. Dairycoates shed had an allocation of 145 engines in 1950, 94 in 1959 and 35 in 1965. 'WD' class 2-8-0s and 'K3' class 2-6-0s predominated in the 1959 list, whilst 'N8' and 'N10' 0-6-2Ts could be seen with other old NER types in the 1950 list. The shed was recoded 50B in 1960.

Hull (Botanic Gardens) (53B, closed 1959) had 50 engines in 1950 and 26 by 1959. The NER opened Botanic Gardens in 1900 and provided two internal roundhouses,. The shed replaced the establishment at Paragon where extensions to the station were subsequently made. In 1959 BR rebuilt the shed for diesel use and converted it into a conventional straight building. 'B1' class 4-6-0s and 'D49' 4-4-0s as well as a solitary Class 3 2-6-0, No 77001, were on the books in 1959.

Hull (Springhead) (53C, closed 1958) was the old Hull & Barnsley eight-road straight shed and housed 55 engines in 1950.

Bridlington (53D, closed 1958) had an allocation of 10 engines, including 'Shire' class 4-4-0s. The shed officially closed in 1958 but remained in use for servicing steam locomotives for several years afterwards. The building was a three-road straight shed erected in 1892 by the NER.

Goole (53E, closed 1967) was of L&YR origin and had 34 engines in 1950, 27 in 1959 and 13 in 1965. Once the haunt of the L&YR 0-6-0STs, the six-road shed became North Eastern Region property in 1956 when the code changed from 25C to 53E. The code changed again in 1960 to 50D.

The grimy 'WD' class 2-8-0 was a familiar sight all over Britain, and Hull (Dairycoates) (53A) had its fair share of these ungainly 'Austerities'. No 90072 is seen on Woodford shed in March 1963. None of the class of 733 BR engines have been preserved, although a Swedish-operated example survives on the Keighley & Worth Valley. *A. N. H. Glover*

Above:
Thompson 'B1' No 61306 is seen at Carnforth in May 1968 awaiting restoration. The class was introduced in 1942 for mixed traffic work on the LNER and No 61306 was built by the North British Locomotive Co in 1947. Of a class of 409 engines in 1959 only two examples have survived, Nos 61306 and 61264. The engine illustrated was shedded at Hull Botanic Gardens (53B) in 1959. *A. N. H. Glover*

Below:
Kielder Viaduct on the Border Counties line sees 'K3' class 2-6-0 on a Newcastle to Hawick train in 1953. No 61897 was allocated to Hull (Dairycoates) (53A) in 1959 and was a member of a large class of Moguls designed by Gresley for light freight and passenger work. *E. E. Smith/Colour-Rail*

55A LEEDS HOLBECK

55A	Leeds (Holbeck)
55B	Stourton
55C	Farnley Junction
55D	Royston
55E	Normanton
55F	Manningham Keighley Ilkley
55G	Huddersfield

Sample Allocation 1959

Leeds Holbeck 55A

Class 3 3-6-2T	40140/93
Class 2P 4-4-0	40491/40552/40690
Class 2 2-6-2T	41267
Class 4 2-6-4T	42052/42138
Class 6P/5F 2-6-0	42771/4/95/8
Class 4 2-6-0	43039/43/55/6/43117/24/30
Class 5 4-6-0	44662/44753–57/44826/8/49/52/3/4/7/ 44943/83/45273/45428
Class 6P/5F 4-6-0	45562/4/5/6/8/9/73/89/97/45605/8/19/ 39/58/9/75/94/45739
Class 7P 4-6-0	46109/12/13/17/45
Class 2 2-6-0	46453/93/8
Class 3F 0-6-0T	47420
Class 8F 2-8-0	48067/48104/57/8/9/48283/99/ 48443/54
Class 7P/6F 4-6-2	70044/53/4
Class 5 4-6-0	73010/45/53/66/9/73171

Total 81 Engines

Holbeck (55A, closed 1967) was the principal Midland Railway shed in Leeds with a twin internal roundhouse in the style of that company. The allocation was 95 engines in 1950, 81 in 1959 and 40 in 1965. Main line power in the 1959 list was represented by 18 'Jubilee' 4-6-0s, five 'Royal Scot' 4-6-0s and three 'Britannia' Pacifics in the 1959 list. A notable resident in the 1965 list was 'Jubilee' No 45593 *Kolhapur*, now preserved. The Holbeck code changed from 20A to 55A in 1957 when the NER took over.

Stourton (55B, closed 1967) was a single roundhouse enclosed in the style of the former Midland Railway. The 1950 allocation was for 48 engines and the 1959 allocation was for 36. By 1965 there were 25 engines allocated. Old MR and LMS classes were the mainstay of the allocation until closure. The code changed from 20B to 55B in 1957 upon the Regional changeover.

Farnley Junction (55C, closed in 1966) was a 12-road shed which originated with the LNWR. It was coded 25G, within the LMR sequence, from 1948 until transfer to the NER when it became 55C — the code it retained until closure. Of its allocation of 50 locomotives in 1950, no less than 23 were ex-WD 2-8-0s, along with more glamorous motive power in the form of five 'Jubilees'. In 1959, there were 22 ex-WD 2-8-0s alongside four 'Jubilees'. Of the latter No 45708 *Resolution* was present on both occasions.

Royston (55D, closed 1967) was an LMS 10-road shed of 1932 with an allocation of 60 locomotives in 1950, 55 in 1959 and 32 in 1965. The locomotives were mainly LMS or MR types with a few 'WD' class 2-8-0s for freight work. The code was changed from 20C in 1957.

Normanton (55E, closed 1967) was of L&YR origin and had an allocation of 49 engines in 1950, 41 in 1959 and 21 in 1965. The locomotive types were mainly of MR or LMS origin with 'WD' class 2-8-0s for freight. The shed was a five-road brick building and the code was changed from 20D to 55E in 1957.

Manningham (55F, closed 1967) was of Midland origin with an enclosed roundhouse. The allocation consisted of 45 engines in 1950, 25 in 1959 and 11 in 1965. The ex-L&YR 2-4-2Ts were in evidence in 1950 and one member of the class, No 50621, has been preserved at the National Railway Museum. The code changed from 20E in 1957.

Huddersfield (55G, closed 1967) was a shed of LNWR origin rebuilt by the LMS and was a traditional LNWR straight building of eight roads. The allocation consisted of 42 engines in 1950, 26 in 1959 and 12 in 1965. 'Jubilee' class No 45596 *Bahamas*, now preserved, was the most famous resident in 1959. The code changed from 25B to 55G upon Regional reorganisation.

Right:
Holbeck (55A) 'Jubilee' No 45562 *Alberta* runs into Appleby West in 1967 with a northbound Settle and Carlisle passenger train. Holbeck had 18 'Jubilees' allocated on the 1959 list, and four members of the class survive. *J. Phillips*

Right:
Holbeck (55A) Standard Class 5 4-6-0 No 73069 double-heading with Class 5 4-6-0 No 45407 can be seen near Blackburn on 4 August 1968 just before steam's demise on BR. *Author*

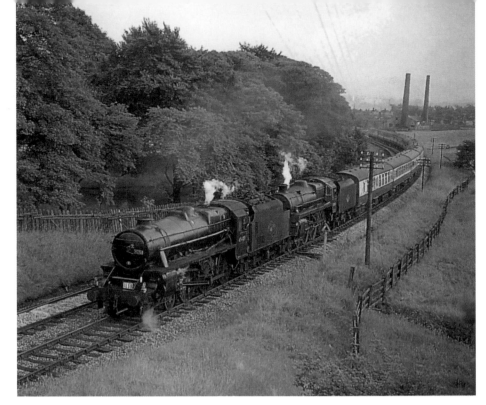

Below:
Class 5 4-6-0 No 45428 of Holbeck (55A) now preserved in LMS livery can be seen shunting BR maroon stock in Cricklewood sidings. The 'Black 5s' were one of Britain's largest class of locomotives and several examples have been preserved. The engine has been named *Eric Treacy* **after the late Bishop of Wakefield.** *Author*

WAKEFIELD

56A	Wakefield
	Knottingley
56B	Ardsley
56C	Copley Hill
56D	Mirfield
56E	Sowerby Bridge
56F	Low Moor
56G	Bradford

Sample Allocation 1950

Wakefield 56A

Class 2MT 2-6-2T	41250–54
Class 5MT 4-6-0	45101/45204–6/9/21/61/45339
Class 2MT 2-6-0	46438/9
Class 3F 0-6-0T	47510/72/3/80/2
Class 8F 2-8-0	48502/4/6/11/14
Class 7F 0-8-0	49625
Class 2P 2-4-2T	50650/6/50712/5/62/4/88/99/50869/ 73/86/92/8
Class 2F 0-6-0ST	51447
Class 2F 0-6-0	52041/3/4
Class 3F 0-6-0	52120/50/4/86/52235/84/52305/19/ 45/51/3/69/86/52433/5/52521/61/76
Class WD 2-8-0	90124/57/63/87/97/90212/37/42/3/9/ 92/90310/29/33/4/7/9/41/2/53/61/2/ 3/70/9/80/1/96/7/90404/12/4/5/7/ 90581/90607/15/7/20/4/31/3/5/7/9/ 43/44/51/2/4/6/67/73/9/82/92/ 90710/9/22/5/9

Total 122 Engines

Wakefield L&YR shed (56A, closed 1967) was a 10-road straight shed and accommodated locomotives for freight work. The allocation in 1950 shows quite a few of the old L&YR types, including the now preserved Barton Wright 0-6-0 No 52044 withdrawn in 1959 and now resident on the Keighley & Worth Valley Railway. The allocation in 1959 consisted of 87 engines and the allocation for 1965 was 84, of which 60 were 'WD' class 2-8-0s. Wakefield shed was completely rebuilt by BR in 1954. The shed was recoded from 25A to 56A in 1956.

Ardsley (56B, closed 1965) was of Great Northern origin and consisted of an eight-road layout. The 1950 allocation was for 88 engines, the 1959 allocation for 63 and the 1965 for 45. Old GNR types gave way to LNER and BR types and eventually in the 1965 allocation the shed had four 'A1' Pacifics on its books as well as two 'V2s' and 17 'B1' 4-6-0s, eight of which were named. Ardsley (rebuilt by BR in 1954) was recoded to 56B from 37A in 1956. The 1950 allocation included an ex-Great Central 'B4' class, No 61482 *Immingham*.

Copley Hill (56C, closed 1964) was the rebuilt GNR shed of five roads. The 1950 allocation was for 39 engines which had been reduced to 33 by 1959. The 1959 list included 10 'A1' class 4-6-2s. The shed was transferred to the North Eastern Region in 1956 when the shedcode was altered from 37B.

Mirfield (56D, closed 1967) was an ex-L&YR shed of eight roads and in 1950 contained 40 engines which became reduced in number to 30 in 1959 and 19 in 1965. LMS and L&YR types gave way to more modern classes after 1950. The shed was recoded in 1956 from 25D. Nine of the Fowler '7F' 0-8-0s of 1929 were allocated in 1950.

Sowerby Bridge (56E, closed 1964) was an old L&YR shed of six roads and was rebuilt by BR in 1954. The allocation was for 33 engines in 1950 and 26 in 1959. The 1950 list was of interest as one Fowler 0-8-0, five L&YR 0-6-0STs and nine L&YR '3F' 0-6-0s were allocated. The shed also had two L&YR 2-4-2Ts, including the rebuilt No 50925. The code changed from 25E in 1956.

Low Moor (56F, closed 1967) was a six-road ex-L&YR shed with 37 engines in 1950, 70 in 1959 and 17 in 1965. In 1959 LMS standard types predominated with a few ex-GNR 0-6-0s; there were also six 'B1s' (three of them named). The code was altered from 25F in 1956.

Bradford Hammerton Street (56G, closed 1958) was a GNR shed of 11 roads. The shed was modified to take diesels in 1954 but the 1950 allocation was for 49 engines, mainly ex-GNR types. The code was changed from 37C in 1956.

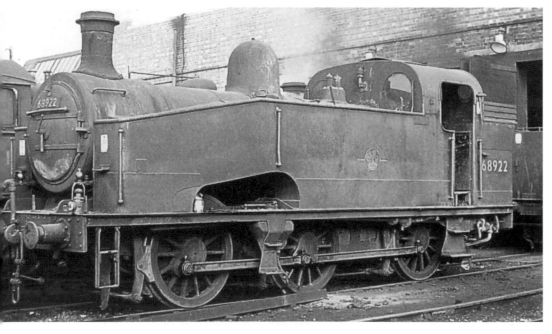

Ardsley shed (56B) was an ex-Great Northern establishment and No 68922 is seen on shed in 1958. The engine was allocated to Low Moor (56F) in 1959 and was a Gresley 0-6-0T Class J50/2 of 1922 of which no examples were preserved. *Colour-Rail*

This Barton Wright 0-6-0 of 1887, No 52044, formerly of
Wakefield (56A), was one of the few L&YR engines to be
preserved. The engine, seen here painted up for a film, was
built by Beyer Peacock & Co and was withdrawn in
1959 *Author*

Wakefield (56A) had a large allocation of 'WD' class 2-8-0s,
there being 60 engines of the class allocated in 1965.
No 90076, in a reasonably clean condition, is seen at Goole
on 8 October 1966 on an LCGB railtour. *A. N. H. Glover*

Scottish Region

INVERNESS

60A	Inverness Dingwall Kyle of Lochalsh
60B	Aviemore Boat of Garten
60C	Helmsdale Dornoch Tain
60D	Wick Thurso
60E	Forres

Sample Allocation 1959
Inverness 60A

Class 5 4-6-0	44718/9/22/3/4/83/4/5/8/9/98/99/ 44991/2/ 45066/90/8/45117/23/4/79/ 92/45319/20/60/1/45453/60/1/76/7/8/9
Class 3P 4-4-0	54463/87/93/6
Class 2P 0-4-4T	55198/9/55216/26/7/36
Class 0F 0-4-0ST	56038
Class 3F 0-6-0T	56300/5/41
Class 3F 0-6-0	57575/94/57661
Total 50 Engines	

The Highland Railway branch to Dornoch was closed to passengers on 13 June 1960. In the final years of operation BR sent two '16XX' class 0-6-0 pannier tanks to replace the ex-HR 0-4-4Ts. Pannier No 1649 is seen in dismal weather waiting to depart with Britain's smallest cathedral in the background. The engine was allocated to Inverness (60A) shed for maintenance purposes. *Author*

Inverness shed was an interesting place and originated from the former Highland Railway. The impressive entrance gateway housed the water tank and was originally designed to be the entrance to a completely circular building. The 1950 allocation of 60 engines included four of the HR 'Clan' goods class 4-6-0s which were withdrawn by the end of 1953. The depot closed in 1962 and was coded as 32A until 1949. Of interest in the 1959 allocation are the ex-Caledonian types which had been re-shedded including the Caley 'Pug' No 56038 used for shed shunting.

Aviemore (60B, closed 1962) had an impressive four-road shed in stone originating from Highland Railway days. The 1950 allocation was for seven engines which was increased to 12 in 1959. Pre-Grouping types included Caley 0-6-0s and 4-4-0s as well as a solitary 'Black 5' 4-6-0. The shed was recoded from 32B in 1949.

Helmsdale (60C, closed 1962) was a two-road wooden affair next to the station platforms. The 1959 allocation of six engines included two ex-Western Region pannier tanks, Nos 1646 and 1649. The engines were examples of a modern class of pannier introduced from 1949 for branch line and shunting work. The two locos worked the Dornoch branch and replaced the ex-Highland 0-4-4Ts in 1957.

Wick (60D, closed 1962) was a two-road stone-built shed with five engines in 1950 and three in 1959. The 1950 allocation included the ex-Highland Railway 'Ben' class 4-4-0s, of which one was earmarked for preservation. No 54398 *Ben Alder* did not make it into preservation and after many years in store was scrapped reputedly because it had a Caledonian boiler.

Forres (60E, closed 1959) had five engines in 1950 and six in 1959, all of ex-Caledonian origin. The shed was a stone structure with a wooden roof with slate tiles and an adjacent foreman's office. The two-road shed was an ideal subject for the model maker.

Above:
Caledonian '3F' 0-6-0 No 57594 is seen at Munlochy in June 1960 with the week's tour of Scotland organised by the RCTS and SLS. The Black Isle branch to Fortrose was closed to passengers on 1 October 1951. Inverness shed (60A) had three Caledonian 0-6-0s in 1959. *Author*

Below:
Inverness (60A) had six ex-Caledonian 0-4-4Ts in 1959. Classified '2P' by the LMS they were used for shunting. One member of the class, No 55189, has been restored to CR blue livery and is owned by the Scottish Railway Preservation Society. *Author*

KITTYBREWSTER

61 A

61A	Kittybrewster
	Ballater
	Fraserburgh
	Inverurie
	Peterhead
61B	Aberdeen (Ferryhill)
61C	Keith
	Banff
	Elgin

Sample Allocation 1950

Kittybrewster 61A

Class 2P 4-4-0	40603/22/50	
Class B1 4-6-0	61134/61307/23/4/43/5/8–52/61400/1/4	
Class B12 4-6-0	61505/7/8/11/13/21/4/6/8/32/39/43/	
	52/60/63	
Class D41 4-4-0	62225/8/9/30/2/41	
Class D40 4-4-0	62260/1/5/8/70/72–9	
Class J36 0-6-0	65247	
Class F4 2-4-2T	67151/7/64	
Class G5 0-4-4T	67287/67327	
Class V1 2-6-2T	67667/71	
Class Z4/Z5 0-4-2T	68190/1/2/3	
Class J72 0-6-0T	68700/10/7/9/49/50	
Class N14 0-6-2T	69125	

Total 70 Engines

Standard Class 4 2-6-0 No 76104 arrives at Craigellachie in July 1959 with the 5.50pm Elgin to Aberdeen. Kittybrewster (61A) had five Class 4s allocated in 1959 and several examples of the class have been preserved. *Author*

The ex-Great North of Scotland Railway half roundhouse at Kittybrewster (61A, closed 1961) held a fascinating selection of old LNER types in 1950 which was replaced by 1959, by more modern types, when the list had been reduced to 52 engines. In 1950 the shed had 15 'B12' 4-6-0s of GER design, two GER 'F4' 2-4-2Ts, ex-North Eastern 'G5s' and 'J72s' as well as an 'N14' 0-6-2T. There were 19 ex-GNSR 4-4-0s and three LMS '2P' 4-4-0s. By 1959 the allocation had changed with 'Glen' class 4-4-0s being drawn in to replace the 'D40/D41' classes. BR Standard Class 4 2-6-4Ts had arrived by 1959 and totalled 11 in number. The GNSR 0-4-2Ts in the 'Z4' and 'Z5' class were still in evidence on the 1959 list but were not to last very long. The 'D40' class 4-4-0 No 62277 *Gordon Highlander* survived into preservation and can now be seen in Glasgow Transport Museum, the only GNSR locomotive to have survived.

Ferryhill (61B, closed 1967) was the principal passenger shed at Aberdeen and was used jointly by the ex-North British and Caledonian Railways. The 1950 allocation shows 40 engines which had declined to 30 by 1959 and 15 by 1965. Star exhibits at Ferryhill were the three 'A2' class 4-6-2s and the eight 'V2' class 2-6-2s which were replaced by eight 'A4' Pacifics after the East Coast main line dieselisation in 1962. Noteable engines allocated to Ferryhill were 'A4s' 60007 *Sir Nigel Gresley*, 60009 *Union of South Africa*, 60019 *Bittern* and 'A2' class No 60532 *Blue Peter*, all of which have been preserved. The shed was a simple 12-road straight affair and could accommodate all engines under cover.

Keith (61C, closed 1961) was a four-road ex-GNSR straight shed in stone with an allocation of 25 engines in 1950 and 22 in 1959. *Glen Douglas*, the now preserved NBR 4-4-0, was allocated here in 1959 along with five 'K2' class 2-6-0s, including two of the named versions.

Above:
Keith (61C) was the home shed for 'D34' class 4-4-0 No 62469 *Glen Douglas* in 1959. The engine was built at Cowlairs in 1913 and was restored by BR in 1959 to North British Railway livery. The engine is now based on the Bo'ness & Kinneil Railway. *Author*

Below:
Kittybrewster (61A) shed had five ex-LMS 4-4-0s in 1959 and one, No 40663, is seen at Forres in 1960 on a special organised by the RCTS and SLS. No 40663 was withdrawn the following year. *Author*

62 A ⬭ **THORNTON**

62A	Thornton
	Anstruther
	Burntisland
	Kirkcaldy
	Ladybank
	Methil
62B	Dundee (Tay Bridge)
	Arbroath
	Montrose
	St Andrews
62C	Dunfermline
	Alloa
	Kelty

Sample Allocation 1959
Thornton 62A

Class 2P 0-4-4T	55217
Class B1 4-6-0	61103/18/33/4/46/7/8/61262/77/ 61330/43/58/61401/3
Class D30 4-4-0	62418
Class D34 4-4-0	62467/75/8/92
Class D11 4-4-0	62677
Class D49 4-4-0	62708/12/6/28/9/33/44
Class J35 0-6-0	64466/74/88/64522
Class J37 0-6-0	64546/9/50/64/5/96/64600/02/16/ 8/29/35
Class J36 0-6-0	65218/52/65345
Class J38 0-6-0	65900–5/7/8/10/11/3/21/5/31/2
Class J88 0-6-0T	68332/4/5/53
Class J83 0-6-0T	68453/6/8/9
Class J72 0-6-0T	69012
Class N15 0-6-2T	69132/43/69223
Class 4 2-6-0	76109/10/11
Class WD 2-8-0	90004/19/20/58/90117/28/68/82/90350/ 90441/72/90513/34/9/90614/90/90705

Total 94 Engines

Thornton (62A, closed 1967) was a traditional straight shed with eight roads and originated from North British Railway days. The 1959 allocation was certainly interesting as there were 16 classes of engine represented in the total allocation of 94. Old NBR 'Scott' and 'Glen' class 4-4-0s were still in use as well as a 'Scottish Director' in the shape of 'D11' class, No 62677 *Edie Ochiltree*. A handful of 'Shire' class 4-4-0s ('D49') a number of NBR 0-6-0s and 17 'WD' class 2-8-0s made for plenty of variety in the allocation.

Dundee Tay Bridge (62B, closed 1967), a six-road straight ex-NBR shed, had 101 engines in the 1950 list and 64 in the 1959 allocation. By 1965 the allocation had come down to 31 engines. With the closure of Dundee West in 1958 all the locomotives from the ex-Caley shed went to Tay Bridge shed nearby. Dundee West was converted into a diesel depot after closure to steam.

Dunfermline (62C, closed 1967) was an NBR four-road shed with 76 locos in 1950, 54 in 1959 and 29 in 1965. In the 1959 list there were 12 classes, mainly ex-NBR types and a few WDs. One of the named 'J36' class 0-6-0s, No 65253 *Joffre*, was allocated there until withdrawal.

The North British 'J88' class 0-6-0Ts were introduced in 1904 and used for dock shunting. No 68346 was an outside cylinder 0-6-0 with wooden buffer beams and is seen at Alloa in 1960. The engine was allocated to Dunfermline shed (62C) in 1959. *Author*

Top:
Dunfermline (62C) with 'D30' class 4-4-0 No 62436 *Lord Glenvarloch* on shed on a Saturday in 1959. The 'Scott' class was introduced by Reid for the NBR in 1912 and consisted of 27 engines. No 62436 was built in 1915 and withdrawn by BR in June 1959. The class was extinct by 1960 when the last two members of the class were withdrawn at St Margarets in June of that year.
A. N. H. Glover

Above:
North British 0-6-0 No 64615 of Dundee Tay Bridge (62B) is seen at Lauriston on the Inverbervie branch in June 1960 with two Caledonian coaches in the train. The 'J37' class 0-6-0s were introduced in 1914 and consisted of a class of 104 engines. No 64615 was built in 1920 and survived until 1963. *Author*

63A PERTH

63A	Perth
	Aberfeldy
	Blair Atholl
	Crieff
63B	Stirling
	Killin
63C	Forfar
63D	Oban
	Ballachulish

Sample Allocation 1950

Perth 63A

Class 4P 4-4-0	40921/2/3/38/9/41125
Class 5MT 2-6-0	42742/3
Class 4F 0-6-0	44193/4/44251/3/4/7/8/44314/8/28
Class 5MT 4-6-0	44698/9/44704/5/96/7/44801/79/85/
	44924/5/31/58–61/72–80/97–9/
	45007/11/86/45118/9/25/7/62/4–7/9/
	70–3/5/45213/66/45309/57/65/6/
	89/45452/6–60/3–7/9/70/2–5/83/8/
	92/6/7
Class 6P/5F 4-6-0	45564/75/45644
Class 3P 4-4-0	54447/8/58/9/67/9/76/85/9/94/9/
	54500–3
Class 2P 0-4-4T	55144/71/5/6/55208/9/12/13/6/8
Class 3F 0-6-0T	56246/90/56328/31/47/52/3/9
Class 2F 0-6-0	57339/45/57450/73
Class D49 4-4-0	62714/25
Class WD 2-8-0	90523/30/90675
Total 138 Engines	

Perth (63A, closed 1967) was a straight eight-road shed of Caledonian Railway origin rebuilt by the LMS. By 1959 the allocation was down to 97 and by 1965, 29 engines. The shed had no less than 75 'Black 5' 4-6-0s on the books in 1950, probably a record for the class. Pacifics were not allowed on to the Highland section and Class 5MT 4-6-0s were the largest engines to be found north of Stanley Junction. Old Caley types were still in use in 1959 for station pilot work and shunts. The shedcode was changed from 29A in 1949 with the change from the old LMS system.

Stirling South (63B, closed 1966) had a total of 49 engines in 1950 and 39 in 1959. The four-road shed in stone of ex-Caledonian vintage closed in June 1960 and was used for storage after closure. The NBR shed at Stirling closed in 1958. The final allocation at 63B was nine 'Black 5' 4-6-0s. The code changed from 31B in 1949 to 63B and then to 65J in 1960.

Forfar (63C, closed 1964) was a four-road shed of CR origin and had 21 engines in 1950. In 1958 the establishment became a sub-shed of Perth. The code changed from 29D to 63C in 1949.

Oban (63D, closed 1962), built by the Caledonian Railway, housed seven engines in 1950, all of ex-CR origin. In 1959 the allocation was the same except that the individual locomotives had changed around. The shed was a two-road affair and the code changed from 31C to 63E in 1949, from 63E to 63D in 1955 and from 63D to 63C in 1959.

'Black 5' class 4-6-0 No 44931 of Perth (63A) approaches Whitemoss crossing with a Perth to Stirling train in June 1963. The engine was built at Crewe in 1946 and was a member of a class of 842 engines built from 1934 to 1951. The engine was withdrawn in 1965. *Author*

Below:
Caley 0-6-0 No 56347 sizzles at Strathord ready to work a society special to Bankfoot in April 1962. The design dated from 1895 and was the standard CR 0-6-0 shunting tank engine classified '3F' by the LMS. Perth (63A) had eight members of the class allocated in 1959. *Author*

Bottom:
Caledonian 4-4-0 No 54485 of the Pickersgill '72' class is seen with the two CR coaches in June 1960 at Perth (63A) on the RCTS/SLS week's railtour. Perth had 15 ex-CR 4-4-0s allocated in 1959 which were classified '3P' by the LMS and BR. *Author*

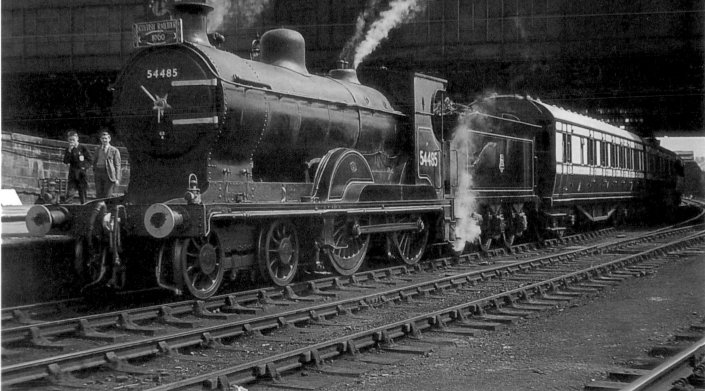

64 A ST MARGARETS

64A	St Margarets (Edinburgh)
	Dunbar
	Galashiels
	Hardengreen
	Longniddry
	North Berwick
	Seafield
	South Leith
64B	Haymarket
64C	Dalry Road
64D	Carstairs
64E	Polmont
64F	Bathgate
64G	Hawick
	Riccarton Junction
	St Boswells

Sample Allocation 1959

Haymarket 64B

Class 2P 0-4-4T	55165
Class A4 4-6-2	60004/9/11/12/24/27/31
Class A3 4-6-2	60035/7/41/3/57/87/9/90/4/6/7/8/9/ 60100/1
Class A1 4-6-2	60152/9/60/1/2
Class A2 4-6-2	60507/9/10/19/29/30/4/5/6/7
Class V2 2-6-2	60816/9/24/7/60920/7/51/7/9
Class B1 4-6-0	61007/76/81/61178/61219/21/44/5
Class D11 4-4-0	62685/90/1–4
Class D49 4-4-0	62705/9/19/43
Class J36 0-6-0	65235/43
Class V1 & V3 2-6-2T	67610/15/20
Class J83 0-6-0T	68457/81
Class N15 0-6-2T	69211
Total 73 Engines	

St Margarets (64A, closed 1967) was of North British origin and consisted of two separate establishments on either side of the running lines. This large establishment had 221 engines in 1950, 175 in 1959 and 40 in 1965. Most NBR types could be seen in 1950 including 'Scott' and 'Glen' classes. In 1965 the shed had acquired two 'A4' and three 'A3' 4-6-2s along with a clutch of BR Standard types. The shed lasted until April 1967. A noticeable survivor was NBR 'Pug' 0-4-0ST No 68095 of 1889, now owned by the Scottish Railway Preservation Society.

Haymarket (64B, closed 1963) was an ex-NBR shed and housed no less than 36 Pacifics in the 1950 allocation of 81 engines. The eight-road shed had 73 engines in 1959, including 37 Pacifics, some of which have become well known such as No 60009 *Union of South Africa* and 60532 *Blue Peter* (in the 1950 list), now both preserved. Another locomotive to be preserved was the 1888-vintage 'J36' class 0-6-0 No 65243 *Maude*, now owned by the Scottish Railway Preservation Society at the Bo'ness & Kinneil Railway. The engine was withdrawn in 1966 and restored to NBR livery.

Dalry Road (64C, closed 1965) was the Caledonian shed in Edinburgh and had 44 engines in 1950, 39 in 1959 and 16 in 1965. A four-road shed, the allocation in 1959 consisted mainly of old CR and LMS types with a few NBR 0-6-0s of the 'J35' and 'J37' classes. The shed was recoded from 28B in 1949.

Carstairs (64D, closed 1966) was an ex-Caledonian four-road shed rebuilt by the LMS with an allocation of 50 engines in 1950, 41 in 1959 and 16 in 1965. Old CR '3P' class 4-4-0s and '2P' 0-4-4Ts as well as '2F/3F' 0-6-0s were still in use in the 1959 allocation. The shedcode changed from 28C to 64D in 1949 and from 64D to 66E in 1960.

Polmont (64E, closed 1964) was a five-road shed of NBR origin with 43 engines in 1950 and 24 in 1959. An interesting selection of old NBR types could be seen in 1959, and out of the 10 classes all were ex-North British including three named 'J36s' and a 'Y9' 0-4-0ST design of 1882, No 68104. The shed was recoded to 65K in 1960.

Bathgate (64F, closed 1966) was originally an NBR six-road shed but was rebuilt by BR into a four-road building in 1954. The 1950 allocation was for 38 engines, the 1959 for 31 and the 1965 for 14. Old NBR types dominated the scene, but BR Standard Class 4MTs and '2MTs' were brought in towards the end.

Hawick (64G, closed 1966), a NBR shed of two roads, had 24 engines in 1950, 16 in 1959 and five in 1965. The 1950 allocation shows eight Class D30 4-4-0s. The Waverley route from Edinburgh to Carlisle, which passed through Hawick, closed on 6 January 1969.

Centre Right:
The Gresley 'A4' Pacifics were allocated to Haymarket (64B) to work East Coast expresses and No 60011 *Empire of India* **was one of the seven 'A4s' on the Haymarket list in 1959. The engine is seen leaving Perth in 1963.** *Author*

Bottom right:
The BR Standard Class 2 2-6-0 was introduced in 1953 and consisted of 65 engines. No 78047 is seen at Coldstream on the former St Boswells to Tweedmouth line in 1962. The engine was allocated to Hawick (64G) in 1959 and was one of two in the class to be shedded at the former Waverley line depot. Out of the 16 engines on the Hawick list, nine different classes were represented. *J. Phillips*

Right:
North British 'J35' class 0-6-0 No 64489 of St Margarets (64A) is seen at Haddington with a railway society special in 1960. The Reid Class B 0-6-0s were built from 1906 until 1913 and consisted of 76 engines. No 64489 was built in 1909 and withdrawn a year after the photograph was taken in June 1961. *Author*

⬭ 65 A EASTFIELD, GLASGOW

65A	Eastfield (Glasgow)
	Arrochar
65B	St Rollox
65C	Parkhead
65D	Dawsholm
	Dumbarton
65E	Kipps
65F	Grangemouth
65G	Yoker
65H	Helensburgh
65I	Balloch
65J	Fort William
	Mallaig

Sample Allocation 1959
Eastfield 65A

Class 4 2-6-0	43135/6/7
Class 5 4-6-0	44702/7/87/44908/56/7/67/8/70/96
Class B1 4-6-0	61140/97/61243/61/61340/2/55/96
Class K2 2-6-0	61764/85–89/94
Class K4 2-6-0	61993–6/98
Class D34 4-4-0	62472/4/7/96
Class D11 4-4-0	62671–76/80/1/2/4/6/7/9
Class J37 0-6-0	64540/1/8/58/78/80/1/64611/22/3/ 32/3/8/9
Class J36 0-6-0	65228/96/65315
Class C15 4-4-2T	67460/74
Class C16 4-4-2T	67485
Class V1 & V3 2-6-2T	67600/2/3/44/64/7/71/80
Class J88 0-6-0T	68345/52
Class J83 0-6-0T	68447/68/79
Class J50 0-6-0T	68952/4–7
Class N15 0-6-2T	69131/63/70/1/8/9/81/2/3/8/91/7/ 69212/4/8
Class 5 4-6-0	73077/8/73105/8/9
Class 4 2-6-0	76074
Class WD 2-8-0	90049/90489

Total 111 Engines

Eastfield (65A, closed 1966) was an ex-North British straight shed of 14 roads with an allocation of 164 engines in 1950 and 111 in 1959. There were 19 different classes in 1959 including seven 'K2s', five 'K4s', four 'D34s' and 13 'D11s'. The most famous resident was the now preserved 'K4' class 2-6-0 *The Great Marquess*, built to work the West Highland line in 1937. One of a class of six, the engine is at present (in 1995) on the Nene Valley Railway, having been withdrawn from BR service in 1961.

St Rollox (65B, closed 1966) was an ex-Caledonian shed with 12 running roads and a two-road repair shed. BR renewed the roof in the 1950s, and the 1959 allocation of 68 engines was mainly of LMS or CR types. Class 5 and 4 Standards were allocated as well as a solitary ex-LNER 'D11' No 62688. The shedcode was changed from 31A to 65B in 1949.

Parkhead (65C, closed 1965) was of NBR origin, and was rebuilt by BR with six roads. The 1950 allocation was for 68 engines and the 1959 for 58. The shed was the stabling point until October 1965 for the four preserved Scottish engines, which were used on railtours.

Dawsholm (65D, closed 1964) was an eight-road CR shed rebuilt by BR in 1949. There were 42 engines in 1950 and 49 in 1959. Engines from the Caledonian were well represented and the 1950 list shows 18 '2F' 0-6-0s as well as Caley 'Pug' No 56029 dating from 1885. The four preserved Scottish engines were kept at Dawsholm until the 1964 closure.

Kipps (65E, closed 1963) of ex-NBR origin was a three-road shed in brick and timber. The allocation in 1950 was for 53 engines and 1959 for 51. Notable inhabitants amongst the ex-NBR stock were the 'Y9' class 0-4-0STs of the wooden tender variety.

Grangemouth (65F, closed 1965) of ex-CR origin was a seven-road dead-end shed with 35 engines in 1950 and 32 in 1959. The best known inhabitant was 'WD' class 2-8-0 No 90732 *Vulcan*, last engine in the class to be built. None of these cheap 'Austerity' class 2-8-0s survived to be preserved from BR, but a Swedish version exists at Haworth. The shed was recoded from 31D to 65F in 1949.

Yoker (65G, closed 1961) had an all-Caley allocation in 1950 with the exception of 'Y9' class 0-4-0ST No 68112. There were 13 engines allocated in 1950 and three in 1959.

Helensburgh (65H, closed 1962) had an allocation of six engines in 1950 and 12 in 1959. The two-road shed of NBR origin was built in brick and became redundant after electrification.

Balloch (65I, closed 1961) was a single-road shed with a corrugated roof built in 1948. The location had been a stabling point for engines since NBR days. The code changed from 65J in 1950.

Fort William (65J) closed in 1962. The 1950 allocation showed 12 engines, including seven named 'K2' 2-6-0s and two 'K4s' as well as three NBR 0-6-0s of the 'J36' class. The 1959 allocation was for 17 engines, including six 'K1' class 2-6-0s. The last steam train ran in 1963 but the shed has been used until recently by visiting 'Black 5s' which work the Mallaig steam service in the summer. The shed was recoded 63D to 65J in 1955 and from 65J to 63B in 1960.

Top:
Eastfield (65A) shed had five 'K4' 2-6-0s in 1959 built by Gresley for working the West Highland line in 1937. There were only six engines in the class and one, No 61994, has been preserved. The engine, now in LNER green livery and numbered 3442 *The Great Marquess*, has worked all over the BR system and is seen here at Brighton in April 1967 having worked a railtour. *J. Phillips*

Above:
St Rollox (65B) 'Black 5' No 45471 is seen on the Dumfries to Stranraer line in 1965 on a freight at Creetown.
No 45471 was built at Crewe in 1938 by the LMS and was withdrawn shortly after the photograph was taken. *Author*

Right:
St Rollox shed (65B) was near the former Caledonian Railway works where 0-4-0ST No 56025 was employed as works shunter. This 1952 photograph depicts the Caley 'Pug' with wooden buffer beam and coloured number plate, the engine is seen in between shunting stints at the works.
T. B. Owen/Colour-Rail

66 A POLMADIE

66A	Polmadie (Glasgow)
66B	Motherwell
66C	Hamilton
66D	Greenock (Ladyburn)

Sample Allocation 1950
Polmadie 66A

Class 4P 4-4-0	40916/41131
Class 4MT 2-6-4T	42167–72/42200-7/13–16/38-47/74–7/88–96/8/9
Class 4F 0-6-0	44196
Class 5MT 4-6-0	44707/87/90/2/3/4/45484–7
Class 6P/5F 4-6-0	45579/83/4/91/2
Class 7P 4-6-0	46102/4/5/7/21
Class 8P 4-6-2	46220-4/27/30/1/2
Class 3F 0-6-0T	47331/2/47536/7/40/1
Class 2P 0-4-4T	55141/67/70/9/97/55201/7/24/8/65/7/8
Class 2F 0-6-0T	56153/4/9/60/2/7
Class 3F 0-6-0T	56239/44/60/1/3/80/1/92/4/5/8/56304–8/14/8/22/4/42/6/9
Class 2F 0-6-0	57230/8/9/68/71/5/88/92/57317/9/20/1/47/60/1/5/7/70/87/8/9/57412/32/3/9/43/4/6/7/8/59/63/4/5
Class 3F 0-6-0	57555/64/81/57619/22/5/30/61/74/90
Total 166 Engines	

Polmadie of Caledonian origin was the principal shed in Glasgow for the West Coast main line and quite a few famous engines were resident. In 1950 there were nine 'Coronation' Pacifics, five 'Royal Scots' and five 'Jubilees'. The 16-road shed closed in 1967, servicing locomotives until the end of steam in Scotland. Famous residents were No 46201 *Princess Elizabeth* and preserved 0-4-4T No 55189 which appear on the 1959 list of 182 engines. Five of the 'Clan' class 4-6-2s were allocated on the 1959 list as well as three 'Britannia' class 4-6-2s. The shed also acquired three 'A2' class 4-6-2s redundant from the East Coast. Polmadie was coded 27A until 1949.

Motherwell (66B, closed 1967) had 116 engines in 1950 and 89 in 1959. The list was reduced to 20 by 1965 but the shed lasted until the end of steam on the Region. The ex-Caley shed of 10 roads was rebuilt by BR in the early 1950s and housed mainly CR or LMS types but 'WD' class 2-10-0s were also shedded there, being 12 in 1950 and nine in 1959. The shed was recoded from 28A in 1949.

Hamilton (66C, closed 1962) was an old Caley shed of 12 roads of ex-CR and -LMS types with an allocation of 51 engines in 1950 and 47 in 1959. The shed was recoded from 27C in 1949.

Greenock Ladyburn (66D, closed 1966) was a six-road shed rebuilt by BR. The allocation was 42 engines in 1950, 39 in 1959 and 12 in 1965. Notable inhabitants were the 'Pug' class 0-4-0STs of the former Caledonian Railway, of which there were three in 1950 and two in 1959. The shed was recoded from 27B in 1949.

Polmadie (66A) was the principal Caledonian shed in Glasgow and the 1959 allocation shows 182 engines of various classes, including main line, passenger and freight. The LMS '8F' class was represented and No 48773, a former 66A engine, is seen double-heading with '5MT' No 44781 near Clitheroe in the last days of steam in 1968. *Author*

Below:
No 46201 *Princess Elizabeth*, **a Polmadie (66A) engine, epitomises main line steam power at Glasgow Central, waiting to work a southbound express over the West Coast main line in 1959.** *J. Swain/Colour-Rail*

Bottom:
'Black 5' No 45029 is seen descending from Greskine signalbox with an up freight on Beattock bank in October 1963. The engine was allocated to 66B Motherwell at the time. *Hugh Ballantyne*

⬭ 67/A CORKERHILL

67A	Corkerhill (Glasgow)
67B	Hurlford Beith Muirkirk
67C	Ayr
67D	Ardrossan
68B	Dumfries
68C	Stranraer Newton Stewart
68D	Beattock

Sample Allocation 1959

Hurlford 67B

Class 3 2-6-2T	40049
Class 2P 4-4-0	40570–3/92/3/7/40605/8/9/12/19/43/4/5/ 40661/5/86–9
Class 6P/5F 2-6-0	42743/4
Class 4F 0-6-0	44281/44312/25
Class 5 4-6-0	45010/45266
Class 2P 0-4-4T	55203/11/64
Class 3F 0-6-0T	56368
Class 2F 0-6-0	57236/84/95/57331/53/83
Class 3F 0-6-0	57562/70/2/7/57637/43/50/1/71/2/89
Class 3 2-6-0	77015–19

Total 55 Engines

Corkerhill (67A, closed 1967) was the GSWR shed for the late lamented St Enoch station. The six-road shed was rebuilt by the LMS and survived until the end of steam in Scotland. The 1950 allocation was for 91 engines and the 1959 for 87. The LMS rid itself of the old GSWR types of loco and brought in old CR types to replace them. By 1959 Standard BR classes had arrived and by 1965 there were 32 engines, all except three being BR Standards. 'Jubilee' class 4-6-0s were also allocated in 1959, there being eight of the class on the list. The shedcode was changed from 30A in 1949.

Hurlford (67B, closed 1966) was a classic GSWR six-road shed in stone with no less than 21 class 2P 4-4-0s for the Ayrshire branch lines in the 1959 allocation as well as five Class 3 77XXX-series 2-6-0s out of a class of 20. Hurlford was recoded from 30B in 1949.

Ayr (67C, closed 1966) was another classic GSWR six-road shed with 59 engines in 1950 and 58 in 1959. By 1965 the allocation had changed to 32 engines, the largest class being 17 'Crab' Class 5 2-6-0s.

Ardrossan (67D, closed 1965) had 35 engines in 1950 and 42 in 1959. LMS '2P' 4-4-0s and CR 0-6-0s were the predominant types in 1959, and Caley No 57566 from this shed is now preserved on the Strathspey Railway in CR blue as No 828. Ardrossan shed was recoded from 30C in 1949.

Kingmoor shed (68A), which was coded 12A from 1948 to 1949 and from 1958 to 1968, was transferred to the London Midland Region from 1958.

Dumfries (68B, closed 1966) was another GSWR six-road shed in stone and had 38 engines in 1950 which were reduced to 33 in number by 1959. Old CR and LMS types could be seen here until replaced by BR Standard types. The code changed from 12G in 1949 and to 67E in 1962.

Stranraer (68C, closed 1966) had two sheds which originated from two different companies that served the town. Both the GSWR and CR sheds were three-road buildings with the 1950 allocation being for 13 engines and the 1959 for 15. By 1965 the allocation was for only three engines. The code changed from 12H to 68C in 1949 and from 68C to 67F in 1962.

Beattock (68D, closed 1967) was a two-road affair in stone with open wooden doors. There were 16 engines allocated in 1950 and 10 in 1959. The shed was provided for locos that banked trains to the summit. The code was changed from 12F in 1949 and to 66F in 1962.

Above:
Kilmarnock pictured in July 1960 with LMS '2P' 4-4-0 No 40661 of Hurlford (67B) shed which in 1959 had 21 engines of the '2P' class allocated out of a total of 55 engines. No 40661 was built at Derby in 1931 to the LMS post-Grouping design which resulted in 138 engines being built. *Author*

Below:
LMS '2P' 4-4-0 No 40665 (67B) built at Derby in 1931 to the Fowler post-Grouping design waits to depart from Kilmarnock in July 1960. Scottish Region blue enamel nameboards and faded paintwork complete the vintage scene. *Author*

Below left:
'Black 5' 4-6-0 No 45432 pauses at the unstaffed station of Crossmichael on the former Portpatrick & Wigtownshire Joint line in May 1965. The engine is working the 8am Stranraer to Dumfries stopping passenger. The engine was one of the four members of the class to be allocated to Dumfries (68B) shed. *Author*

Southern Region

70 A NINE ELMS

70A	Nine Elms
70B	Feltham
70C	Guildford
70D	Basingstoke
70E	Reading South
70F	Fratton
70H	Ryde (Isle of Wight)

Sample Allocation 1959
Nine Elms 70A

Class 57XX 0-6-0PT	4634/72/86/92/8/9770
Class M7 0-4-4T	30123/33/30241/5/8/9/30319/20/1
Class T9 4-4-0	30338/30718/9
Class N15 4-6-0	30457/30763/74/8/9
Class H15 4-6-0	30482/4/6/9/91/30521–4
Class 700 0-6-0	30694/9/30701
Class V 4-4-0	30902/3/6/7
Class U 2-6-0	31617/21/4/34/31796
Class E4 0-6-2T	32487/97/8/32500/63
Class Q1 0-6-0	33015/17/38
Class WC/BB 4-6-2	34006/7/9/10/8/20/9/31/47/64/5/ 90/3/4/5
Class MN 4-6-2	35005/12/4/6–20/9/30
Class 5 4-6-0	73087/8/9/73110–19

Total 90 Engines

Nine Elms (70A, closed 1967) consisted of two sheds side by side: the 'old shed' of 15 roads (1889) and 'new shed' of 10 roads (1910). There had been an engine shed on the site since the railway first reached Waterloo, the shed being the principal LSWR shed and having an allocation of 200 engines before the Grouping. It was an obvious target for the Luftwaffe and suffered severe damage from 1940 to 1943. The engine allocation for 1950 was for 99 engines which by 1965 had fallen to 39 examples. The shed was one of the last to be in use on the Region and closed in July 1967. Locomotives allocated in 1959 include ex-GWR pannier tanks as well as old LSWR, SR and BR Standard types. Five Bulleid Pacifics shown on the list have been preserved including No 35029 *Ellerman Lines* displayed in sectionalised form at the National Railway Museum.

Feltham (70B, closed 1967) was the LSWR London area main freight shed with an allocation of 77 engines in 1950, 60 in 1959 and nine in 1965. The concrete structure opened in 1923 and was of six roads. The shed housed some rare freight classes, notably the Urie 'H16' 4-6-2Ts (five in class) and the 'G16' 4-8-0Ts (three in class).

Guildford (70C, closed 1967) was unusual for the Region as being a roundhouse and dated from 1887. The 1950 allocation was for 57, the 1959 for 45, and the 1965 for 28 engines. The locomotives included LSWR and SR freight and mixed traffic types. A well-known inhabitant until 1954 was the LSWR 0-4-0ST *Ironside* of 1890 built by Hawthorn Leslie for Southampton Docks.

Basingstoke (70D, closed 1967) was an LSWR shed of three roads dating from 1905. In 1950 there were 21 engines on the list which by 1959 had been reduced to 14. The shed was the home for the seven 'Remembrance' class 4-6-0s, the Maunsell rebuild of the old Brighton Baltic tanks withdrawn in 1957.

Reading South (70E, closed 1965) was a three-road shed of SECR origin with an allocation of 17 engines in 1950. The shed lost its coded status in 1959.

Fratton shed was jointly owned between the LSWR amd LBSCR. The shed was of the enclosed roundhouse type and used for storage after closure in 1959. The shed was recoded from 71D to 70F in 1954.

Ryde (70H, closed 1967) was originally a two-road shed of Isle of Wight Railway origin, in timber. The 1959 allocation was for 21 engines, all 'O2' class 0-4-4Ts except for two ex-LBSCR 'E1' class 0-6-0Ts. One locomotive has survived from this allocation — 'O2' class No W24 *Calbourne*, which can now be seen at work on the Isle of Wight Steam Railway. The shedcode was changed from 71F in 1954. The building was a two-road affair, having been constructed by the SR in 1930 to replace the IWR building.

Top:
Rebuilt Bulleid No 34093 *Saunton* **heads westward with the 08.30 Waterloo to Bournemouth near Wimbledon in June 1967, a few days before the elimination of steam on the Region. No 34093 was a Nine Elms (70A) engine at the time.** *J. Phillips*

Above:
'King Arthur' class 4-6-0 No 30457 *Sir Bedivere* **shunts empty stock at Basingstoke in December 1960 with LSWR lower quadrant signals in the background. The 'King Arthur' class totalled 74 engines and one, No 777** *Sir Lamiel***, has been preserved to work specials. No 30457 was a Nine Elms (70A) engine in 1959.** *J. Phillips*

Left:
Feltham (70B) was the LSWR main freight shed for the London area and the 'S15' class 4-6-0 heavy freight locomotive was very much in evidence at the depot. Both variants, the Urie and the Maunsell, were shedded at Feltham. No 30500 is seen on Basingstoke shed in December 1960. *J. Phillips*

71 A EASTLEIGH

71A	Eastleigh
	Andover Junction
	Lymington
	Southampton Terminus
	Winchester City
71B	Bournemouth Central
	Branksome
71G	Weymouth
	Bridport
71I	Southampton Docks

Sample Allocation 1959
Eastleigh 71A

Class M7 0-4-4T	30028/9/30/33/30125/30/
	30328/75/7/8/9/30479–81
Class B4 0-4-0T	30083/88/96
Class T9 4-4-0	30117/20/30287/8/9/30300
Class O2 0-4-4T	30212/23/9
Class 700 0-6-0	30306/16
Class H15 4-6-0	30473–77
Class Q 0-6-0	30530/1/2/5/6/42/3
Class N15 4-6-0	30770/3/84/5/6/8/9/90/1
Class LN 4-6-0	30850–9/30861–3
Class U 2-6-0	31618/9/20/9/39/92–5/31801/2/3/8
Class E4 0-6-2T	32491/32510/56/9/79
Class Q1 0-6-0	33020/1/3
Class 3 2-6-2T	41293/41305
Class 4 2-6-0	76010–9/76025–9/76063–9
Class 3 2-6-2T	82012/4/5/6
Total 111 Engines	

Eastleigh (71A, closed 1967) was a 15-road shed built by the LSWR in 1903 to cater for traffic for nearby Southampton Docks. The 1950 allocation was for 142 engines, with the 1965 list showing 103. The 1959 list shows 15 different classes including old LSWR, LBSCR and SR types as well as BR Standard classes. Several locos from the 1959 list have been preserved, notably 'T9' No 30120 and 'Lord Nelson' No 30850, now in the National Collection. Other preserved locos are 'U' class 2-6-0 No 31618 and BR Standard Class 4 No 76017, the former being resident on the Bluebell Railway and the latter on the Mid Hants Railway. Eastleigh had several sub-sheds including Winchester with its solitary ex-LSWR 'B4' class 0-4-0T. The Eastleigh code was changed to 70D in 1963.

Bournemouth (71B, closed 1967) was a four-road shed rebuilt prewar by the Southern Railway. In 1950 there were 52 engines, which had increased to 60 by 1960 but had declined to 39 by 1965. Bournemouth had 16 'M7' class 0-4-4Ts in 1959 for local branches but the main traffic was the express passenger service to Waterloo. The 1959 list shows seven 'King Arthur' 4-6-0s, three 'Lord Nelson' 4-6-0s and 20 Bulleid Pacifics, of which eight have been saved for preservation. The shedcode was changed to 70F in 1963.

Weymouth (71G, closed 1967) was the old GWR shed which was recoded from 82F in 1958 and recoded again as 70G in 1963. The 1950 allocation shows 31 engines, all of GWR origin, but by 1959 the number had changed to 26 with BR Standard types taking over. The Southern Region had assumed responsibility in 1958 and by 1965 the list consisted of 12 'Merchant Navy' class 4-6-2s, five Class 2 2-6-2Ts and six Class 5 Standard 4-6-0s. Four of the Bulleids have been saved for preservation.

Southampton Docks (71I, closed 1967), a three-road shed of LSWR origin, had 14 0-6-0Ts in the 'USA' class in 1959 and three 'E2' class 0-6-0Ts. The code was altered from 71I to 70I in 1963. Four of the USA 0-6-0Ts have been preserved.

The 'Lord Nelson' class four-cylinder 4-6-0 consisted of only 16 engines introduced in 1926 by the Southern Railway under Maunsell. No 30860 *Lord Hawke*, a Bournemouth (71B) engine, is seen 'on shed' at Nine Elms in March 1962. *J. Phillips*

Above:
Bulleid's unrebuilt Pacifics, known to footplatemen as 'Spams' after the well-known tinned meat, could kick up quite a cloud of smoke at times. 'West Country' class 4-6-2 No 34105 *Swanage*, now resident on the Mid Hants Railway, is seen near Reading West in July 1963 with an inter-Regional train heading south. The engine was shedded at Bournemouth (71B) in 1959. *Author*

Below:
Drummond 'M7s' worked branch lines in Hampshire and No 30111, a Bournemouth (71B) engine, can be seen at Holmsley with the push and pull train from Brockenhurst. Two examples of the class have survived into preservation. *Author*

⬭ 72 A ⬭ EXMOUTH JUNCTION

72A	Exmouth Junction
	Bude
	Callington
	Exmouth
	Lyme Regis
	Okehampton
	Seaton
72B	Salisbury
72C	Yeovil Town
72E	Barnstaple Junction
	Ilfracombe
	Torrington
72F	Wadebridge

Sample Allocation 1959
Exmouth Junction 72A

Class M7 0-4-4T	30021/3/4/5/7/44/5/30323/74/
	30667–70/76
Class O2 0-4-4T	30182/30199/30232
Classs 700 0-6-0	30317/27/30691
Class 0415 4-4-2T	30582–4
Class T9 4-4-0	30702/9/11/5/7/26
Class S15 4-6-0	30841–6
Class Z 0-8-0T	30950/3/5/6/7
Class U 2-6-0	31790/1
Class N 2-6-0	31830–47/49
Class EIR 0-6-0T	32697
Class WC/BB 4-6-2	34002/11/5/23/4/30/2–6/8/56/7/8/
	60/1/2/3/9/72/4/5/6/9/80/1/96/
	34104/6/8/9/10
Class MN 4-6-2	35003/8/9/11/3/23/6
Class 2 2-6-2T	41306/7/18
Class 3 2-6-2T	82010/1/3/7/8/9/22–5

Total 115 Engines

Exmouth Junction (72A, closed 1965) was an 11-road shed reconstructed by the SR in the 1920s. The LSWR had had a shed on the site since 1887 and this was the principal shed in the West of England for that company. The code was changed from 72A to 83D in 1963 following the Western Region's takeover. The 1950 allocation was for 123 engines and the final allocation was for 23. Some famous engines were allocated here, notably the '0415' class 4-4-2T No 30583 and 'West Country' Pacific No 34023 *Blackmore Vale*, both now resident on the Bluebell Railway. Five Bulleid Pacifics have been preserved out of the shed's total of 40, one of the largest Pacific allocations on BR.

Salisbury (72B, closed 1967) was a 10-road straight shed dating from LSWR days with an allocation of 57 engines in 1950, 47 in 1959 and 24 in 1965. The 1959 list included six 'King Arthurs' and 11 Bulleids including No 34051 *Winston Churchill*, now in store at the NRM.

Yeovil Town (72C, closed 1965) of ex-LSWR origin was a three-road shed with 15 engines in 1950 and 10 in 1959. The 1959 allocation included a mixture of ex-GWR and -SR engines as the GWR shed at Pen Mill closed in 1959. The Western Region took over in 1962 and the code was changed from 72C to 83E in 1963.

Barnstaple (72E, closed 1964) was a two-road shed adjacent to Barnstaple Junction. It originated with the LSWR and was transferred to Western Region control in December 1962. It was not recoded — to 83F — officially until September 1963. It closed the following September.

Wadebridge (72F, closed 1964) of ex-LSWR ownership was a two-road affair adjacent to the station. The allocation for 1950 and 1959 was for five engines, of which two were '0O2' class 0-4-4Ts and three were '0298' class 2-4-0 well tanks. Two of the famous Beattie 1874 well tanks — Nos 30585 and 30587 — have been preserved. The shed's code changed to 84E in 1963 as a result of the Western Region takeover.

Below left:
The rebuilding of the Bulleid Pacifics represented one of the better locomotive reconstructions on BR and the 'Merchant Navy' class 4-6-2s lasted until the end of steam on the Region. No 35003 *Royal Mail* is seen at Earlsfield in 1967 with the 8.30am ex-Waterloo. The engine was an Exmouth Junction (72A) locomotive in 1959. *J. Phillips*

Above:
Worting Junction is where the West of England main line diverged from the Bournemouth route. No 34008 is seen taking the flyover on 10 June 1967 with the 9am Weymouth to Waterloo shortly before the demise of steam on the Region. No 34008 *Padstow* was an Exmouth Junction engine (72A) in 1959. *Author*

Below:
Eastleigh in October 1963 sees an unadvertised workman's train hauled by an Ivatt Class 2 2-6-2T No 41295. The engine was allocated to Barnstaple Junction (72E) in 1959 where Ivatt's Class 2s had been sent to replace ageing Drummond tanks. *J. Phillips*

73 A ⬭ STEWARTS LANE

73A	Stewarts Lane
73B	Bricklayers Arms
73C	Hither Green
73D	Gillingham (Kent)
73E	Faversham
73F	Ashford (Kent)
73G	Ramsgate
73H	Dover Folkestone
73J	Tonbridge

Sample Allocation 1959

Stewarts Lane 73A

Class N15 4-6-0	30767/8/9/93/5/30802/3
Class V 4-4-0	30908/9/15/37/8/9
Class E1 4-4-0	31019/67
Class O1 0-6-0	31048/31370
Class D1 4-4-0	31145/31545/31743/9
Class H 0-4-4T	31261/5/31550/1/2
Class C 0-6-0	31317/31575/8/81/3/4/31719/24
Class N 2-6-0	31408–14/31810–2
Class P 0-6-0T	31558
Class U1 2-6-0	31894/5/7/8/31904–7
Class W 2-6-4T	31914/15/21
Class E2 0-6-0T	32100/2/3/6
Class C2X 0-6-0	32543/7
Class WC/BB 4-6-2	34066/7/8/77/85–9/91/2/34100/1
Class MN 4-6-2	35001/15/28
Class 2 2-6-2T	41290/1/2
Class 4 2-6-4T	42087–91
Class 5 4-6-0	73041/2/80–6
Class 4 4-6-0	75074

Total 96 Engines

Stewarts Lane (73A, closed 1963) was the ex-SECR shed which served Victoria and consisted of 16 roads. The shed was rebuilt by the SR in 1934 with an allocation of 170 engines, including locomotives from the LBSCR Battersea shed. The allocation list of 19 classes reflects the considerable variety in the work performed by the engines, from express passenger to ordinary goods. Old SECR and LBSCR types mingled with modern Pacifics and Standards. Some famous locomotives on the 1959 list were No 35028 *Clan Line* and 34092 *City of Wells*, both now preserved Bulleid Pacifics. The shed was recoded to 75D in 1962. Prior to the 1934 rebuild the shed was known as 'Longhedge' or 'Battersea'.

Bricklayers Arms (73B, closed 1962) was the South Eastern Railway's London shed and dated from the earliest years. The old shed consisted of 10 roads and the newer St Patricks shed consisted of four. There were also workshops leading off from a turntable at the rear of the old shed. The Luftwaffe used the buildings as a run-in target for the City of London and the already cramped site became difficult to use. The allocation in 1950 totalled 140 engines and in 1959, 94. Locos consisted of old SECR and SR classes with 13 'Schools' class 4-4-0s and seven Bulleids. The establishment was probably unique as a motive power depot in that it was named after a pub situated in the Old Kent Road.

Hither Green (73C, closed 1961) was the principal freight shed in southeast London. The 1950 allocation was 51 and the 1959 allocation was for 45 engines. The shed had one passenger duty, for which a 'King Arthur' class 4-6-0 was allocated. The six-road shed was opened in 1933 by the Southern to serve the nearby marshalling yard.

Gillingham (73D, closed 1960) had 38 engines in 1950 and 24 in 1959. The building was a three-road brick structure and housed mainly SECR types.

Faversham (73E, closed 1959) was a four-road building of SECR origin which was rebuilt by BR to accommodate diesels. The allocation in 1950 was for 31 engines which was reduced to 26 in 1959.

Ashford (73F, closed 1963) was formerly coded 74A and a principal shed until the Kent Coast electrification. An allocation of 63 engines was reduced to 43 in 1959. The 10-road shed, opened in 1931, was of concrete and replaced the former four-road building next to the works. Old SECR types prevailed including the unique 'J' class 0-6-4Ts and Wainwright's 'S' class 0-6-0ST No 31685 converted from a 'C' class 0-6-0.

Ramsgate (73G, closed 1960) was built by the SR in 1930 as part of the reorganisation of lines in East Kent. The allocation in 1950 was for 43 engines, which declined to 39 in 1959. The six-road shed had 12 'Schools' class 4-4-0s as well as nine Bulleid 4-6-2s on the 1959 list. The shed was recoded from 74B in 1958.

Dover (73H, closed 1961) was a five-road shed built by the SR in 1928 to replace the former SECR premises at Dover Priory. The shed had 68 locos in 1950 and 61 in 1959. In 1959 there were six 'N15' class 4-6-0s including the now preserved No 30777 *Sir Lamiel* and seven Bulleid 4-6-2s. The shed was recoded from 74C in 1958.

Tonbridge (73J, closed 1965) was a six-road shed dating from SECR times but had been re-roofed in 1952. The allocation was 61 engines in 1950 and 45 in 1959. The code was changed from 74D in 1958.

Right:
The workhorse of the former SECR, the 'C' class 0-6-0, could be found anywhere on the old South Eastern & Chatham system. 'C' class No 31690 in ex-works condition can be seen at Stewarts Lane (73A) in June 1960. One example of the class has survived to be preserved. *Author*

Right:
'Schools' class No 30912 *Downside* is seen on Stewarts Lane loco in 1960. The engine is unusually clean and must have been polished up for a special, possibly a royal for Derby day. The locomotive was allocated to Bricklayers Arms (73B) in 1959 and worked passenger trains on the Hastings line. *Author*

Below:
The SECR rebuilds were a very successful design and popular with the enginemen. No 31739, a 'D1' class 4-4-0, is seen on Ashford shed in March 1961 shortly before steam finished in the area. The engine was a Bricklayers Arms (73B) loco in 1959. None of the class was preserved. *Author*

75 A BRIGHTON

75A	Brighton
	Newhaven
75B	Redhill
75C	Norwood Junction
75D	Horsham
75E	Three Bridges
75F	Tunbridge Wells West

Sample Allocation 1950
Brighton 75A

Class P 0-6-0T	31178/31556
Class U1 2-6-0	31890–4
Class I1X 4-4-2T	32005/32595
Class H1 4-4-2	32037/8/9
Class I3 4-4-2T	32076/86/8
Class E1 0-6-0T	32139/42/32689
Class K 2-6-0	32337/9/41–7/9
Class D3 0-4-4T	32368/72/6/86/93
Class E5 0-6-2T	32400/32573/5/6/83/94
Class H2 4-4-2	32421/2/4/5/6
Class C2X 0-6-0	32437/8/43/32523/8/39/43
Class E3 0-6-2T	32455
Class E4 0-6-2T	32470/1/5/86/94/32504/5/8/9/13/4/66/77
Class A1X 0-6-0T	32636/47
Class WC 4-6-2	34036–41
Class WD 2-8-0	90247/90354
Total 75 Engines	

Brighton (75A, closed 1964) was a cramped nine-road shed which dated from 1861. The shed was rebuilt by the SR in 1938 and as the 1950 sample shows was crammed full of old 'Brighton' types, from the diminutive 'A1X' class 0-6-0Ts to large boilered Atlantics (out-shedded at Newhaven). The allocation even showed some 'WD' class 2-8-0s but was reduced to 62 by 1959. Some famous survivors include the two 'P' class 0-6-0Ts, 'A1X' No 32636 (No 72 *Fenchurch*) and Bulleid 4-6-2 No 34039 *Boscastle*, which is now on the Great Central Railway at Loughborough.

Redhill (75B, closed 1965) had 31 engines in 1950 and 24 in 1959. The three-road shed of SECR origin was rebuilt by the SR in 1928 and was inhabited by old SECR and LBSCR classes until they were ousted by Standard BR types. The 1965 allocation shows 18 Class 4MT 2-6-4Ts and a solitary 'N' class 2-6-0. One of the Class 4 tanks, No 80151, has been saved for preservation.

Norwood Junction (75C, closed 1964) was a five-road building erected by the SR in 1935 and contained freight types. The allocation in 1950 was for 38 and in 1959, 23 engines. A well-known engine allocated in the 1950 list is No 32473, the 'E4' class now on the Bluebell Railway restored to original condition as No 473 *Birch Grove*.

Horsham (75D, closed 1964) was unusual in that the shed of 1896 vintage was an open roundhouse of a type rare in Britain. The allocation in 1950 was for 25 engines and in 1959 for 15.

Three Bridges (75E, closed 1964) was a three-road shed dating from 1909 with an allocation of 31 engines in 1950 and 28 in 1959. 'Q' class No 30541, a resident on the 1950 list, is now preserved on the nearby Bluebell Railway.

Tunbridge Wells West (75F, closed 1965) was an important place in the heyday of the LBSCR but the allocation for 1950 was for 24 engines, which was reduced to 22 in 1959. The four-road shed was reroofed by the SR after bomb damage. The LBSCR 'J1' and 'J2' class 4-6-2Ts Nos 32325 and 32326 were allocated here in 1950.

The LBSCR 'K' class 2-6-0s were introduced by L. B. Billington in 1913 and a most successful class of Mogul they were too. The class was allocated to Brighton (75A), Three Bridges (75E) and Fratton (70F) in 1959 but regrettably no example was preserved. One of the class is seen at Purley as standby for the Royal Train in June 1962. *J. Phillips*

Above:
Old LBSCR engines survived until the 1960s and the 'E4' class 0-6-2Ts lasted until 1963. No 32470 can be seen at Petworth with the Midhurst goods in 1962. The engine was allocated to Horsham shed (75D), which was a roundhouse. *J. Phillips*

Left:
SECR 'H' class No 31530 in ex-works condition can be seen at Rowfant in February 1962 on the push and pull to East Grinstead. The engine was allocated to Three Bridges (75E) at the time and one member of the class, No 31263, has been preserved on the nearby Bluebell line.
J. Phillips

Western Region

OLD OAK COMMON

81 A

81A	Old Oak Common
81B	Slough Marlow
81C	Southall
81D	Reading Henley-on-Thames
81E	Didcot
81F	Oxford Fairford

Sample Allocation 1959
Old Oak Common 81A

Class 15XX 0-6-0PT	1500/3/4/5
Class 2251 0-6-0	2222/76/82
Class 57XX 0-6-0PT	3648/88/3754/4615/44/5717/64/ 7722/34/91/8751/3/4/6/7/9/60–5/7– 73/9658/9/61/9700–7/09/10/25/51/ 4/8/84
Class 4073 4-6-0	4082/90/6/5008/14/27/34/5/40/3/ 4/52/6/60/5/6/74/82/4/7/93/ 7001/4/8/10/3/7/20/4/5/7/30/2/3/6
Class 47XX 2-8-0	4700/1/2/4/8
Class 49XX 4-6-0	4900/19/5923/9/31/2/6/9/40/1/54/8/ 5976/87/6920/42/59/61/2/6/73/4/8/ 90/7902/3/4/27
Class 60XX 4-6-0	6000/2/3/9/12/3/5/8/9/22/3/4/8
Class 61XX 2-6-2T	6110/1/3/20/1/32/5/41/2/4/5/9/58/ 9/68
Class 94XX 0-6-0PT	8434/59/9400/10/1/2/4/6/8/9/20/3/79
Class 9F 2-10-0	92229/30/8/9/40/1/4–7
Total 173 Engines	

Old Oak Common (81A, closed 1965) was the Great Western's largest shed and the principal depot for the London area. The building dated from 1906 and had a quadruple turntable enclosed roundhouse accommodating a very large allocation of express passenger locomotives which included 13 'Kings', 35 'Castles', 28 'Halls' and five '47XX' class Churchward 2-8-0s. Some famous engines on the 1959 list include the three now preserved 'Kings' Nos 6000, 6023 and 6024.

Slough (81B, closed 1964) was a GWR four-road shed with a single-road extension. The allocation for 1950 was for 48 engines, which was reduced to 36 in 1959. All of the locomotives were tank engines for suburban and branch line work. A notable survivor is '61XX' class No 6106, now at the GWS Didcot.

Southall (81C, closed 1965) was an eight-road building,

having been rebuilt in 1954 from the original six-road shed of GWR origin. The allocation for this freight shed was for 71 engines in 1950 and 58 in 1959. By 1965 there were 42 engines on the list with ex-GWR freight classes being supplemented by BR Class 9 2-10-0s.

Reading (81D, closed 1965) was a nine-road shed dating from 1930 when the GWR rebuilt the shed from the 1880 roundhouse. There were 91 engines in 1950 and 75 in 1959. Five 'Castles' and 26 'Halls' were allocated in 1959. Two famous occupants in 1950 were the ex-MSWJR 2-4-0s Nos 1335 and 1336.

Didcot (81E, closed 1965) was a four-road shed with 47 engines in 1950 and 48 in 1959. The 1965 allocation shows 11 'Halls' and three 'Manors' including No 7816 *Frilsham Manor*, the last engine to run on BR in pre-Nationalisation livery. The shed still exists as the Great Western Society's headquarters and working museum.

Oxford (81F, closed 1966) had 54 engines in 1950, 62 in 1959 and 26 in 1965. The four-road building with timber roof had 11 'Halls' and four 'Granges' allocated in 1965 and was the last ex-GWR shed with steam on the Western Region when it closed in January 1966.

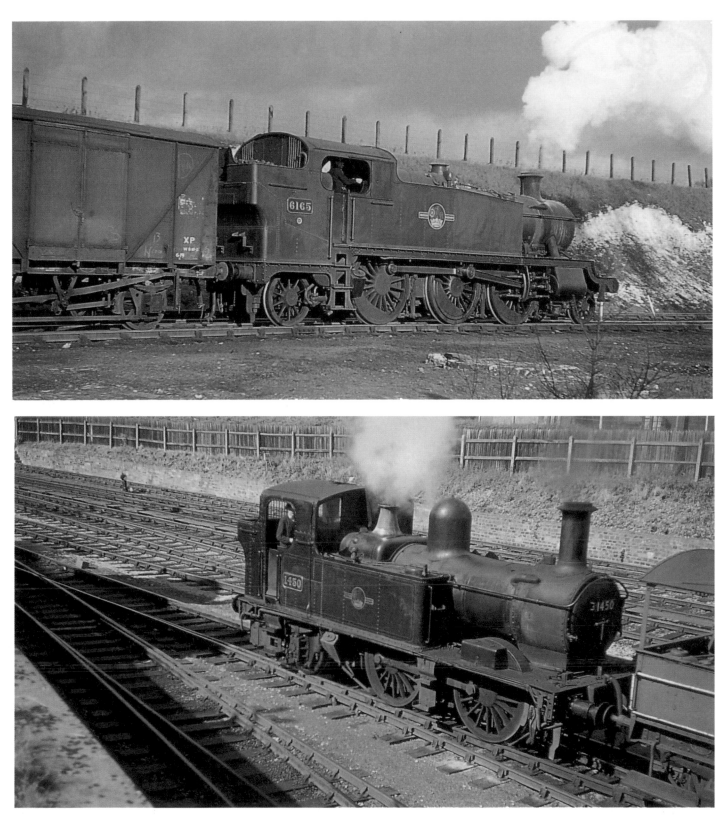

Left:
Magnificent 'Castle' class 4-6-0 No 5018 *St Mawes Castle* can be seen 'on shed' at Reading (81D) in 1963. The lined Brunswick green looks a little grubby but many locos became very run-down towards the end of steam on BR. *Author*

Top:
A large Prairie tank of Southall (81C) is seen shunting at Basingstoke in April 1964. No 6165 is depicted here in unlined GWR green with BR emblem on the tank sides. One member of the class, No 6106, has been preserved by the GWS at Didcot. *J. Phillips*

Above:
Collett 0-4-2T No 1450 formerly of Slough (81B) shed in faded Brunswick green is one of the '14XX' class engines to have been preserved. The class was introduced in 1932 to work branch lines on the former GWR system. *J. Phillips*

82A	Bristol (Bath Road)
	Bath
	Wells
	Weston-super-Mare
	Yatton
82B	St Philip's Marsh
82C	Swindon
	Chippenham
82D	Westbury
	Frome
82E	Bristol (Barrow Road)
82F	Bath (Green Park)
	Radstock West
	Highbridge East
82G	Templecombe

Sample Allocation 1959
Bristol Bath Road 82A

Class 10XX 4-6-0	1000/5/9/11/4/24/8
Class 14XX 0-4-2T	1409/12/54/63
Class 57XX 0-6-0PT	3677/3720/48/59/4619/8741/ 9623/6/9/71
Class 4073 4-6-0	4075/8/9/80/1/5015/48/54/7/62/73/ 6/8/85/90/2/6/7003/11/4/5/8/9/34
Class 51XX 2-6-2T	4163/5186/97
Class 49XX 4-6-0	4922/7/47/88/5949/50/6900/8/19/ 54/7/72/81/2/97/7901/7
Class 43XX 2-6-0	5311/6363
Class 45XX 2-6-2T	5529/30/6/61
Class 61XX 2-6-2T	6107
Class 94XX 0-6-0PT	9481/8
Class 2 2-6-2T	41202/3/49
Class 3 2-6-2T	82007/33/5/40/2/3/4
Total 84 Engines	

Bristol Bath Road (82A, closed 1960) was a 10-road shed rebuilt by the GWR in the 1930s. The site had been used since earliest times and had included a double enclosed roundhouse. The 1950 allocation was for 100 engines, including No 6000 *King George V*, but by 1959 the list had been reduced to 84 engines, including a few standard BR engines as well as seven 'County' class 4-6-0s.

Bristol St Philip's Marsh (82B, closed 1964) was of the enclosed roundhouse type with 141 engines in 1950 and 112 in 1959. The allocation included 60 0-6-0 pannier tanks in 1950 with 23 'Halls' and 17 'Granges'.

Swindon (82C, closed 1964) had 125 engines in 1950 and 108 in 1959, the facilities consisting of a nine-road straight shed and enclosed roundhouse. The 1950 allocation was for 125 engines, which by 1959 had come down to 108. The 1959 list included three 'County' class 4-6-0s and 11 'Halls'. The last 'County' worked from Swindon in November 1964 but the most famous resident in 1959 was *City of Truro* which was kept for specials and railtours.

Westbury (82D, closed 1965) was a five-road shed with an allocation of 75 engines in 1950 and 53 in 1959. 'Halls', of which there were 13 examples allocated, and pannier tanks were the mainstay of the shed's list in 1959.

Bristol Barrow Road (82E, closed 1965) was the former Midland Railway shed having been recoded from 22A in 1958 when the Western Region took it over from the London Midland Region. The shed was an MR enclosed roundhouse in brick and had 58 engines in 1950, 53 in 1959 and 41 in 1965. GWR types superseded the old MR and LMS classes after 1958. 'Jubilee' No 45699 *Galatea* which was purchased for spare parts and is now destined for permanent preservation as well as 'Hall' class 4-6-0 No 4920 *Dumbleton Hall* were both at Barrow Road in 1959.

Bath Green Park (82F, closed 1966) was the Somerset & Dorset shed recoded from 22C to 71G in 1950 and from 71G to 82F in 1958. The allocation in 1950 was for 51, 1959 for 50 and 1965 for 19 engines. The well-known '7F' 2-8-0s were shedded here and two of them, Nos 53808 and 53809, have been preserved.

Templecombe (82G, closed 1966), of the former SDJR, was a two-road shed in brick and concrete with 17 engines in 1950, 18 in 1959 and 13 in 1965. The building had replaced a wooden structure in 1951 and was recoded 22D to 71H in 1950, from 71H to 82G in 1958 and from 82G to 83G in 1963. The old MR types disappeared under WR ownership from 1958 onwards.

Centre right:
Faringdon in April 1959 sees an enthusiasts' special worked by GWR 0-6-0ST No 1365 of the rare Churchward saddletank class used for dock shunting. No 1365 was a Swindon (82C) engine in 1959, and another member of the class, No 1363, has since been preserved by the GWS. *G. H. Hunt/Colour-Rail*

Bottom right:
GWR 2-6-0 No 6327 of St Philip's Marsh (82B) is seen at Andover Junction in September 1961. The Churchward Moguls were introduced in 1911 and in 1959 there were still 169 in stock out of the original 342 which were built from 1911 to 1932. *Author*

Right:
Mogul No 6309 and 'Castle' class No 5038 *Morlais Castle* **can be seen on Reading shed in July 1963. No 6309 was a Swindon (82C) engine in 1959. The 'Castle' class has a straight-sided tender.** *Author*

NEWTON ABBOT

83A	Newton Abbot
	Kingsbridge
83B	Taunton
	Bridgwater
83C	Exeter
	Tiverton Junction
83D	Laira (Plymouth)
	Launceston
83E	St Blazey
	Bodmin
	Moorswater
83F	Truro
83G	Penzance
	Helston
	St Ives
83H	Plymouth (Friary)

Sample Allocation 1959
Newton Abbot 83A

Class 14XX 0-4-2T	1452/66/70
Class 16XX 0-6-0PT	1608
Class 28XX 2-8-0	2805/7/46/75/81/3834/40/41/64
Class 57XX 0-6-0PT	3659/3796/9633/78
Class 4073 4-6-0	4037/83/4/98/5003/11/24/32/49/
	53/5/79/7000/29
Class 51XX 2-6-2T	4105/8/45/50/74/6–9/
	5150/3/4/8/64/78/83/95/6
Class 45XX 2-6-2T	4561/5558/73
Class 49XX 4-6-0	4905/20/36/55/67/75/
	5920/67/6933/8/40/7916
Class 43XX 2-6-0	6360
Class 56XX 0-6-2T	6614
Class 68XX 4-6-0	6813/29/36/59
Class 74XX 0-6-0PT	7445
Class 94XX 0-6-0PT	9440/62/87
Total 74 Engines	

Newton Abbot (83A, closed 1963) was a seven-road shed next door to the works and an important centre in South Devon where a shed and works had existed since South Devon Railway days. The allocation in 1950 consisted of 73 engines and in 1959, 74. 'Granges', 'Castles' and 'Halls' were allocated, a famous resident on the 1950 list being the then brand-new and now preserved No 7029 *Clun Castle*.

Taunton (83B, closed 1964) was a GWR enclosed roundhouse with 58 engines in 1950 and 56 in 1959. A relic in 1959 was the ex-Cardiff Railway 0-4-0ST No 1338. 'Hall' class No 4930 *Hagley Hall*, now preserved on the Severn Valley Railway, was also a resident in 1959.

Exeter (83C, closed in 1963) was a four-road shed which had 35 engines in 1950 and 27 in 1959. The 1959 allocation included five '14XX' class 0-4-2Ts used for local branch lines.

Plymouth Laira (83D, closed 1964) was a four-road shed and an enclosed roundhouse with an allocation of 108 engines in 1950 and 81 in 1959. Famous residents in 1950 included No 7027 *Thornbury Castle*, now awaiting restoration, 0-6-0ST No 1363 and, in the 1959 list, No 7812 *Erlestoke Manor*. There are nine preserved 'Manor' class 4-6-0s. The site of Exeter's steam shed has now been converted into a diesel depot.

St Blazey (83E, closed 1962) was a half roundhouse in brick with end opening doors to the different roads. The allocation for 1950 was 32 engines which increased to 37 in 1959. A well-known resident in 1959 was No 7816 *Frilsham Manor*, still in GWR livery.

Truro (83F, closed 1962) replaced the old Cornwall Railway shed in 1900 and was later rebuilt to became a seven-road shed and workshop under BR. The allocation for 1950 was for 23 engines and 1959 for 27.

Penzance (83G, closed 1962) was a five-road shed with a healthy allocation of GWR 4-6-0s including 'Granges', 'Halls' and 'Counties'. The allocation in 1950 was 30 and in 1959, 34 engines. No 'Grange' class 4-6-0, of which Penzance had 12 in 1959, has been preserved.

Plymouth Friary (83H, closed 1963) was the old LSWR shed of three roads which the Western Region acquired in 1958 when the code was altered from 72D. The 1950 allocation was for 23 engines which was reduced to 11 by 1959. Two well-known engines on the 1950 list were the ex-Plymouth Devonport & South Western Junction Railway 0-6-2Ts Nos 30757 and 30758, *Earl of Mount Edgcumbe* and *Lord St Levan*.

Above right:
'Castle' class 4-6-0 No 7029 *Clun Castle* was a Newton Abbot (83A) engine in 1959 and has since been preserved. The engine was fitted with a double-chimney by BR. In 1959 there were 163 members of this famous class in service. *Author*

Centre right:
The Hawksworth 'County' class was introduced in 1945 and was a powerful design of express 4-6-0 classified by BR as '6MT'. Some of the class were fitted with double-chimneys and No 1002 *County of Berkshire* can be seen leaving Truro in 1958 with the down 'Cornish Riviera Express'. The engine was shedded at Penzance (83G) at the time. *J. Phillips*

Bottom right:
GWR 2-6-2T No 4117 is seen at Tiverton Junction with a stopping train to Taunton in the summer of 1958. The engine was based at Exeter (83C) and was a member of the '5101' class, a Collett development of the Churchward 2-6-2T of 1903. Several examples of the class survive on various preserved railways. *J. Phillips*

84 A WOLVERHAMPTON STAFFORD ROAD

84A	Wolverhampton (Stafford Road)
84B	Oxley
84C	Banbury
84D	Leamington Spa
84E	Tyseley Stratford-upon-Avon
84F	Stourbridge Junction
84G	Shrewsbury Clee Hill Craven Arms Knighton Builth Road
84H	Wellington (Salop)
84J	Croes Newydd Bala Penmaenpool Trawsfynydd
84K	Wrexham (Rhosddu)

Sample Allocation 1959
Wolverhampton Stafford Road 84A

Class 57XX 0-6-0PT	3615/64/3756/78/92/8726/96/8
Class 51XX 2-6-2T	4161/5151/87
Class 49XX 4-6-0	4901/86/90/5900/26/6975
Class 4073 4-6-0	5019/22/6/31/45/6/7/59/63/70/2/ 88/9/7026
Class 60XX 4-6-0	6001/5/6/8/11/4/7/20
Class 64XX 0-6-0PT	6418/22
Class 94XX 0-6-0PT	8411/25/6/9428/35/96

Total 47 Engines

Wolverhampton Stafford Road (84A, closed 1963) was a six-road straight shed and two enclosed roundhouses — a GWR rebuild of earlier structures. The allocation for 1950 consisted of 66 and in 1959 47 engines. Express passenger locomotives were the mainstay of the allocation in 1959, a notable survivor to this day being the 4-6-0 No 5900 *Hinderton Hall*.

Wolverhampton Oxley (84B, closed 1967) was of the GWR enclosed roundhouse type with 67 engines allocated in 1950, 51 in 1959 and 58 in 1965. The shed was taken over by the LMR in 1963 and became coded 2B. Two 'Manor' class 4-6-0s have survived from the 1965 list: No 7820 *Dinmore Manor* and No 7821 *Ditcheat Manor*.

Banbury (84C, closed 1966) was a four-road standard Churchward shed of 1908 with an allocation of 70 in 1950, 52 in 1959 and 23 in 1965. The LMR recoded the shed to 2D in 1963. GWR standard types were supplemented by Standard Class 9F 2-10-0s under LMR auspices.

Leamington Spa (84D, closed 1965) was a standard four-road shed dating from 1906 with 29 engines in 1950, 20 in 1959 and 13 in 1965. The LMR took over in 1963 and recoded the shed to 2L. LMR types were introduced after 1963.

Tyseley (84E, closed 1966) was a GWR enclosed roundhouse with workshops and still sees steam today as the Birmingham Railway Museum. In 1950 there were 118 engines, in 1959 70, and in 1965 39. The 1950 list includes 40 0-6-0 pannier tanks and 27 '51XX' class 2-6-2Ts.

Stourbridge (84F, closed 1966), recoded to 2C when the LMR took over, had a 1950 allocation of 85, a 1959 allocation of 64 and a 1965 allocation of 33 engines.

Shrewsbury (84G, closed 1967) consisted of two sheds side by side, a GWR roundhouse and a LNWR straight nine-road shed. The Western Region ran the shed from 1949 to 1963 and the LMR from 1963 to 1967. The codes changed from 4A to 84G to 1949, from 84G to 89A in 1960 and from 89A to 6D in 1963. The allocation in 1950 was 120, 1959 105 and 1965 45 engines. Famous residents include six now preserved 'Manors': Nos 7802, 7812, 7819, 7822, 7827 and 7828 out of the seven allocated in 1965!

Wellington (84H, closed 1964) was a three-road ex-GWR shed in brick with a timber roof. The allocation in 1950 was for 24 engines which was reduced to 15 by 1959. The code was changed to 2M in 1963.

Croes Newydd (84J, closed 1967) was of the enclosed roundhouse type with 54 engines in 1950, 41 in 1959 and 38 in 1965. The shed was recoded to 89B in 1960 and to 6C in 1963. LMR types were shedded after 1963.

Wrexham Rhosddu (84K, closed 1960) was the old GCR shed which came into the Western Region in 1958 when the code was changed from 6E. Old GCR types dominated the 1950 list of 29 engines which had been reduced in 1959 when ex-GWR types were drawn in. The shed was used for storage until 1964.

Right:
The GWR Collett 0-6-0 design was introduced in 1930 and was used on light passenger trains and freight working. No 3200, a Shrewsbury (84G) engine in 1959, is seen leaving Pencader in 1962 bound for Aberystwyth. Only one member of the class has been preserved, No 3205. *Author*

Above:
**'Hall' class No 6930 *Aldersey Hall*
of Stourbridge Junction (84F) is
seen at Basingstoke with a fitted
freight in the summer of 1963.
Several 'Hall' class 4-6-0s have
been preserved and work special
trains from time to time.** *J. Phillip*

Right:
**'King' class No 6014 *King Henry
VII* thunders northwards near
Princes Risborough in 1962 with
an express for Wolverhampton.
Three members of the class have
been preserved and two have been
restored to working order.
No 6014 was a Wolverhampton
(84A) engine in 1959.** *Author*

85A	Worcester
	Evesham
	Kingham
85B	Gloucester
	Brimscombe
	Cheltenham (Malvern Road)
	Cirencester
	Lydney
	Tetbury
85C	Hereford
	Leominster
	Ross
85D	Kidderminster
85E	Gloucester (Barnwood)
	Dursley
	Tewkesbury
85F	Bromsgrove
	Redditch

Sample Allocation 1959

Worcester 85A

Class 16XX 0-6-0PT	1629/61
Class 2251 0-6-0	2206/9/43/7/3205/13/4/6/7/8
Class 57XX 0-6-0PT	3605/7/3725/75/4613/4/25/9/64/ 7707/77/8737
Class 4073 4-6-0	4088/9/5029/37/42/71/81/7005/7
Class 51XX 2-6-2	4109/13/24/42/52/4/5179
Class 49XX 4-6-0	4952/5917/5952/6/71/80/4/94/ 6947/8/50/84/7920/8
Class 68XX 4-6-0	6807/20/51/6/77
Class 43XX 2-6-0	7319/38
Class 81XX 2-6-2T	8106
Class 94XX 0-6-0PT	8427/80/96/9401/29/55/66/80/6
Class 4 4-6-0	75003/25
Class 2 2-6-0	78001/8/9
Class 3 2-6-2T	82008/30/8
Total 79 Engines	

Worcester (85A, closed 1965) had two shed buildings, one with four roads and the other with three. The allocation in 1950 was 81 engines, in 1959 79 engines and in 1965, 24 engines. A well-known resident in 1959 was the (now) sole surviving Collett 0-6-0, No 3205, built in 1946, which has been preserved in working order, as well as the last engine built by the GWR, No 7007 *Great Western*.

Gloucester (85B, closed 1965) was a GWR standard straight building with 11 roads. There were 99 engines allocated in 1950, 84 in 1959 and 35 in 1965. Well-known residents were No 7808 *Cookham Manor* and, in 1965, No 7029 *Clun Castle*, now both restored.

Hereford (85C, closed 1964) was a GWR straight shed of eight roads. The allocation in 1959 was for 53 engines, which was reduced to 35 by 1959.

Kidderminster (85D, closed 1964) was a two-road shed of 19 engines in 1950 and 15 in 1959. The Cleobury Mortimer & Ditton Priors Railway 0-6-0PTs Nos 28 and 29 appeared in the 1950 allocation although they were of course sub-shedded out on to the light railway. The code changed to 84G in 1960 and to 2P in 1963 upon the LMR takeover.

Gloucester Barnwood (85E, closed 1964) was the Midland Railway shed in the city with 42 engines in 1950 and 31 in 1959. The shed was of the MR enclosed roundhouse type and contained old MR and LMS types including the 0-4-0Ts of Deeley vintage dating from 1907 (Nos 41535 and 41537). The code changed from 22B to 85E in 1958 and from 85E to 85C in 1961.

Bromsgrove (85F, closed 1964) was a three-road shed of MR origin with an allocation of 10 engines in 1950 and 11 in 1959. The allocation for 1950 shows the MR 0-10-0 No 58100 and the LNER Garratt 2-8-8-2T No 69999 kept for banking the famous Lickey incline. The shedcode changed from 21C to 85F in 1958 and from 85F to 85D in 1960. No 58100, known as 'Big Bertha', was withdrawn in 1956.

A rural scene on the Western Region in 1961 with ex-GWR 0-4-2T No 1445 at Easton Court on the now closed Woofferton to Tenbury Wells branch of the former LNWR/GWR joint line. The engine was shedded at Hereford (85C) on the 1959 allocation list. *J. Phillips*

Above:
GWR 2-6-2T No 5514 is seen at Bourton-on-the-Water in July 1959 with the 7.14pm Cheltenham to Kingham. The engine was shedded at Gloucester (85B) at the time and was a member of the '4575' class, a Collett modification of the Churchward design of 1906. A total of 101 members of the '45XX/75' class were in use in 1959 and several examples have been preserved. *Author*

Right:
Tewkesbury engine shed was of Midland Railway origin and ex-MR 0-6-0 No 43754 of Gloucester Barnwood (85E) can be seen on the right with pannier No 4614 (85A) on the left inside the shed. The shed was a sub-shed of Gloucester Barnwood. *J. Phillips*

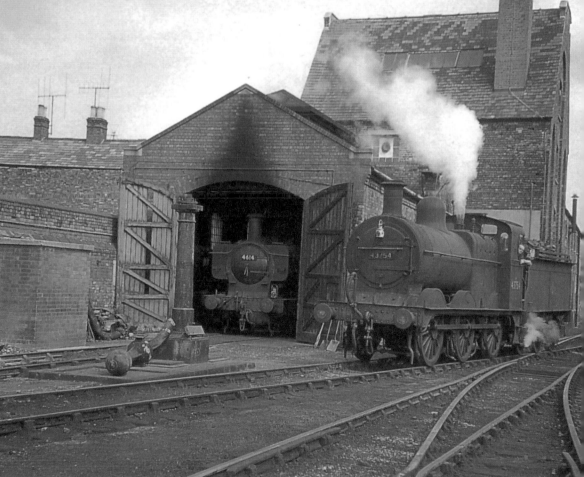

86A	Newport (Ebbw Junction)
86B	Newport (Pill)
86C	Cardiff (Canton)
86D	Llantrisant
86E	Severn Tunnel Junction
86F	Tondu
86G	Pontypool Road
86H	Aberbeeg
86J	Aberdare
86K	Tredegar

Sample Allocation 1959
Cardiff Canton 86C

Class 15XX 0-6-0PT	1508
Class 28XX 2-8-0	2834/64/74/7/89/91/5/3801/9/10/6/7/35/42/3/5/55/60
Class 57XX 0-6-0PT	3670/3755/4622/33/5727/49/76/7775/8723/8/8776/9603/48/9713/23/59/78
Class 4073 4-6-0	4073/5095/99/7023
Class 42XX 2-8-0T	4207/25/31/54/66/70/1/97/5207/18/60/1
Class 49XX 4-6-0	4946/56/73/99/5910/1/46/62/70/6901/32/5/6/9/43/58/63/99/7913
Class 56XX 0-6-2T	5602/85/6600
Class 43XX 2-6-0	6308/26/33/52/7332
Class 72XX 2-8-2T	7227
Class 78XX 4-6-0	7805
Class 94XX 0-6-0PT	8439/41/7/57/64/6/84/9426/37/43/53/61/77/93/4
Class 7P/6F 4-6-2	70016/8/9/20/2–9
Class WD 2-8-0	90125/88/90201/38/90312/23/90524/65/72/3/9/90685/91/3
Class 9F 2-10-0	92003/5/92231–37

Total 131 Engines

Newport Ebbw Junction (86A, closed 1965) was one of the enclosed roundhouse type sheds with two turntables and workshops. The shed replaced the High Street establishment which was demolished in 1916. A large allocation of GWR freight types could be seen as the coal traffic from the valleys was extensive, the allocation for 1950 being for 141 engines, 1959 for 127 and 1965 for 27 engines. The code was changed from 86A to 86B in 1963.

Newport Pill (86B, closed 1963) was the shed formerly owned by the Alexandra Docks & Railway Co and was a two-road building in brick. The 1950 allocation was for 57 and the 1959 for 50 engines. On the 1950 list there were two ex-ADR 0-6-0Ts, Nos 666 and 667. The 1959 list consisted of 32 pannier tanks and 18 '42XX' 2-8-0Ts.

Cardiff Canton (86C, closed 1962) was the principal passenger shed in the district and was recoded to 88A in 1960. The shed was of the enclosed roundhouse type with a six-road straight building adjoining. 'Castle' class No 4073 *Caerphilly Castle* was the most famous resident in 1959 as well as the 12 'Britannias' and nine '9F' 2-10-0s.

Llantrisant (86D, closed 1964) was a GWR three-road shed with 19 engines in 1950 and 16 in 1959. The sole remaining ADR 2-6-2T No 1205 was on the 1950 list. The shed was recoded to 88G in 1960.

Severn Tunnel Junction (86E, closed 1965) was a six-road standard GWR shed with 75 engines in 1950, 73 in 1959 and 12 in 1965. Twenty-five 2-8-0s in the '28XX' class were allocated in 1959.

Tondu (86F, closed 1964) was a GWR enclosed roundhouse situated in the triangle near the station, with 48 locos in 1950 and 45 in 1959. The allocation consisted of tank engines only. The code was changed to 88H in 1960.

Pontypool Road (86G, closed 1965) was an eight-road building with 88 engines in 1950 and 1959. GWR standard goods types were the mainstay of the shed's allocation in 1959 including six 'Grange' class 4-6-0s.

Aberbeeg (86H, closed 1964) was a four-road shed of the GWR standard type with 37 engines in 1950 and 34 in 1959. The allocation was for tank engines only and the code was altered to 86F in 1960.

Aberdare (86J, closed 1965) was an enclosed roundhouse with 52 engines in 1950 and 49 in 1959. The shed was recoded 88J in 1960 and consisted of ex-GWR freight types only.

Tredegar (86K, closed 1960) was an ex-LNWR four-road shed and was coded as 4E until 1949. The 1959 allocation was for 10 engines which included two of the old LNWR 0-8-0s of the G2 class.

Centre right:
Cardiff Canton (86C) plays host to Class 9F No 92220 *Evening Star* in April 1960. The engine spent some time in South Wales and is well known for being the last steam engine to be built by BR. The engine works specials from time to time. *Author*

Bottom right:
GWR pannier tank No 3617 of Llantrisant (86D) does a spot of shunting in Llantrisant Yard in the summer of 1960. The huge class of tank engines built for goods work and shunting was introduced from 1929 onwards and known as the '57XX class, of which there were 779 examples in 1959. *Author*

Above right:
Cardiff Canton (86C) had the Western Region's allocation of 'Britannia' class Pacifics. No 70022 *Tornado* is seen on Canton shed on a Sunday morning in 1960. Two examples of the class survive and are steamed regularly. *Author*

87 / A NEATH

87A	Neath
	Glyn Neath
	Neath (N&B)
87B	Duffryn Yard
87C	Danygraig
87D	Swansea East Dock
87E	Landore
87F	Llanelly
	Burry Port
	Pantyffynnon
87G	Carmarthen
87H	Neyland
	Cardigan
	Milford Haven
	Pembroke Dock
	Whitland
87J	Goodwick
87K	Swansea (Victoria)
	Gurnos
	Llandovery
	Upper Bank

Sample Allocation 1959
Landore 87E

Class 2251 0-6-0	2226/84
Class 28XX 2-8-0	2821/44/3849
Class 57XX 0-6-0	3678/3701/13/68/85/97/8788/9/94/ 9637/9715/38/75/7
Class 4073 4-6-0	4074/6/93/4/7/9/5004/13/6/39/41/ 5051/77/91/7002/9/12/6/28/35
Class 51XX 2-6-2T	4106/7
Class 49XX 4-6-0	4910/23/37/5913/55/88/90/ 6905/12/8
Class 56XX 0-6-2T	5631/56/73/6649/80/8/95
Class 72XX 2-8-2T	7200/7/9/17/24/30/6
Class 94XX 0-6-0PT	8463/9436/84

Total 68 Engines

Neath (87A, closed 1965) had an allocation of 57 engines in 1950 and 63 in 1959; the shed was an enclosed roundhouse with two turntables. The all-tank allocation included the ex-Burry Port & Gwendraeth Valley Railway 0-6-0ST No 2192 *Ashburnham.*

Duffryn Yard (87B, closed 1964) was a six-road shed originating from the Port Talbot Railway and modernised by the GWR. The allocation in 1950 was for 54 engines and 1959 for 58. The locomotives were all tank classes with pannier 0-6-0s being in the majority.

Danygraig (87C, closed 1960) was a stone-built shed of six roads with a three-road workshop of the former Rhondda & Swansea Bay Railway. There were 33 locos in 1950 and 30 in 1959 with some rare classes allocated. All the engines were tanks used for shunting with '1101' class 0-4-0Ts, Swansea Harbour Trust 0-4-0STs and a Powlesland & Mason 0-4-0ST No 1151.

Swansea East Dock (87D, closed 1964) had 30 engines in 1950 and 33 in 1959. The allocation consisted of tank engines except for a solitary '43XX' class No 5361. The shed, dating from 1893, was a three-road building.

Landore (87E, closed 1961) was the principal passenger shed in the area and consisted of two sheds of four and five roads each. There were 60 engines in 1950 and 68 in 1959. Famous engines at Landore included No 4003 *Lode Star,* now in Swindon Museum, and No 5051 *Earl Bathurst,* now at Didcot.

Llanelly (87F, closed 1965) had 86 engines in 1950, 82 in 1959 and 22 in 1965 and consisted of an enclosed roundhouse with two boarded turntables opened by the GWR in 1925. Freight engines occupied the shed and included 0-6-0 tanks from the former Burry Port & Gwendraeth Valley Railway in the 1950 list.

Carmarthen (87G, closed 1964) was a GWR straight shed with workshops and 43 engines in 1950 and 1959. 'Halls', 'Castles' and 'Manors' could be seen at the shed in 1959 including the 'Castle' No 5080 *Defiant,* currently awaiting restoration.

Neyland (87H, closed 1963) was a two-road shed with an allocation of 45 engines in 1950 and 46 in 1959. Four 'County' class 4-6-0s were allocated in 1959.

Whitland (87H, closed 1963) was a sub-shed of Neyland until 1963 when Neyland closed and the 87H code was adopted. The single-road shed of corrugated iron had 15 engines allocated in 1963 which included two 'Manors'.

Goodwick (87J, closed 1963), with an allocation of 14 engines in 1950 and 16 in 1959, was a two-road shed of GWR standard design.

Swansea Victoria (87K, closed 1959) was the LNWR shed which in 1950 had 48 engines and in 1959, 55. The shed was a six-road building and housed ex-LMS types including 19 Stanier '8F' 2-8-0s used on freight trains on the Central Wales line.

Centre right:
Drysllwyn in August 1962 was on the former LNWR line from Carmarthen to Llandilo. Pannier No 7437 is in charge of the two-coach branch train on the line which closed on 9 September 1963. No 7439 was the sole representative of the class at Danygraig (87C) in 1959. *Author*

Bottom right:
Little Mill Junction was the point where the Monmouth branch diverged from the main line from Pontypool. A view from the busy platform in the summer of 1960 shows 'County' class No 1029 *County of Worcester* heading north with an express for Shrewsbury. The engine was a Neyland (87H) engine on the 1959 list. *Author*

Right:
Carmarthen in steam days with pannier No 7439 on the right and Collett 0-6-0 No 3200 of the '2251' class on the left ready to work an Aberystwyth train. The pannier tank No 7439 of the '74XX' class was based at Danygraig (87C) in 1959. *Author*

CARDIFF RADYR

88 A

88A	Cardiff (Radyr)
	Cathays
88B	Cardiff East Dock
88C	Barry
88D	Merthyr
	Dowlais Cae Harris
	Dowlais Central
	Rhymney
88E	Abercynon
88F	Treherbert
	Ferndale

Sample Allocation 1959

Cardiff Radyr 88A

Class 94XX 0-6-0PT	3401–9/8420/38/55/60/9/70/1/8/81
Class 57XX 0-6-0PT	3672/3727/9679
Class 51XX 2-6-2T	4143/60
Class 56XX 0-6-2T	5640/8/63/9/75/83/92/6603/6/7/8/
	12/8/24/6/33/5/8/47/8/59/60/5/
	82/4/9/99
Class 64XX 0-6-0PT	6411/34
Class 72XX 2-8-2T	7202/5/43
Total 55 Engines	

Cardiff Radyr shed had an all-tank allocation and was closed in 1965 having been recoded to 88B in 1960. The 0-6-2T type was the ideal engine for working the valley lines and the '56XX' class is predominant on the allocation list. The shed was of the GWR standard type of four roads and was opened in 1931.

Cardiff East Dock (88B, closed 1965) was an eight-road shed dating from 1931 and had an allocation of 'Castles', 'Halls', 'Granges' and 'Manors' after Canton closed to steam in 1962. The 1950 allocation shows 62, the 1962, 68 and the 1965, 30 engines. The shed closed from 1958 to 1962 and was coded 88B to 1958, 88L from 1962 to 1963 and 88A from 1963 until closure. 'Manor' class No 7820 *Dinmore Manor* was resident on the 1962 list.

Barry (88C, closed 1964) was an ex-Barry Railway shed of six roads with an all-tank allocation of 80 engines in 1950 and 31 in 1959.

Merthyr (88D, closed 1964) was a three-road shed in brick with a corrugated roof, 55 engines in 1950 and 34 in 1959.

Abercynon (88E, closed 1964) was a two-road shed dating from 1929, with 27 engines in 1950 and 25 in 1959 all tanks.

Treherbert (88F, closed 1965) was a GWR shed of 1931 with four roads and 22 engines in 1959. The allocation consisted of

Rhymney in April 1960 shows a clean '56XX' class 0-6-2T 'on shed'. Fully lined out and with BR crest the engine was one of a class of 200 Collett engines designed to work the steep Welsh valleys. The engine was shedded at Merthyr (88D) in 1959. *Author*

Above:
Pannier tank No 9631 shunts empty stock at Merthyr in November 1961. The engine was one of the '57XX' class standard 0-6-0 pannier tank designed by Collett for shunting and light freight work. The engine was shedded at Merthyr (88D) in 1959. *J. Phillips*

Below:
Rural scene in Wales with '64XX' class 0-6-0 pannier at Cefn Coed with the 9.28am Merthyr to Pontsticill Junction in 1961. No 6416 was a Merthyr (88D) engine in 1959 and the line was closed on 13 November 1961. *J. Phillips*

⬭89 A OSWESTRY

89A	Oswestry
	Llanidloes
	Moat Lane
89B	Brecon
89C	Machynlleth
	Aberayron
	Aberystwyth
	Portmadoc
	Pwllheli

Sample Allocation 1959
Oswestry 89A

Class 14XX 0-4-2T	1432/58
Class 16XX 0-6-0PT	1602/3/4
Class 2251 0-6-0	2219/36/9/75/94/3200/2/8/9
Class 57XX 0-6-0PT	3600/3789/5726/9681
Class 54XX 0-6-0PT	5400/22
Class 43XX 2-6-0	6342
Class 64XX 0-6-0PT	6404
Class 74XX 0-6-0PT	7405/10/34
Class 78XX 4-6-0	7800/1/7/19/22/7
Class 90XX 4-4-0	9005
Class 2 2-6-0	46503/4/5/7/9–15/23/4/6/7
Total 47 Engines	

Oswestry (89A, closed 1965), of Cambrian Railways origin, was a six-road shed rebuilt by BR. The 1950 allocation was for 53 engines with the 1959 list being for 47. The shed was recoded 89D in 1960 and 6E in 1963 when the LMR took over. Famous residents in 1959 were 'Manors' Nos 7819, 7822 and 7827, which are now all preserved.

Brecon (89B, closed 1962) had an allocation of 11 engines in 1950 and 13 in 1959. The two-road shed became a sub-shed of Oswestry in November 1959. The shed was recoded to 88K in 1961.

Machynlleth (89C, closed 1966) was of Cambrian origin and had three roads. The building was in brick and modernised by the GWR. The allocation in 1950 was for 53 engines and in 1959 47, which by 1965 had been reduced to 13. The LMR took over in 1963 and the code was changed from 89C to 6F. The 1959 allocation included the now-preserved 'Manor' class No 7802 *Bradley Manor* as well as two elderly 'Dukedog' class 4-4-0s Nos 9015 and 9017.

The restored 'Manor' class 4-6-0 No 7819 *Hinton Manor* rests at Machynlleth after working trips to Aberystwyth and Pwllheli. The engine was allocated to Oswestry (89A) in 1959 and has been restored to BR lined black livery. Nine members of the class have been preserved. *Author*

Above:
The Mid Wales line from Moat Lane Junction to Three Cocks Junction was closed on 31 December 1962. The line was the haunt of the Ivatt 2-6-0s and No 46510 is seen leaving Boughrood & Llyswen shortly before closure. The engine was based at Oswestry (89A) on the 1959 allocation list. *J. Phillips*

Below:
Boughrood & Llyswen in the summer of 1962 with 2-6-0 No 46508 on a freight. The engine was based at Brecon (89B) and the station is now a council storage depot. Some parts of the platform can still be seen. *Author*

Index